Valorization of Resources from Urban Mined Materials

World Scientific Series on Advances in Environmental Pollution Management

Series Editors: Dharmendra Kumar Gupta
(Ministry of Environment, Forest and Climate Change, India)
Clemens Walther
(Leibniz University Hannover, Germany)

Published

Vol. 3   *Valorization of Resources from Urban Mined Materials*
edited by Dharmendra K Gupta, Pankaj Pathak & Dheeraj Mittal

Vol. 2   *Radionuclide Uptake in Food and Consequences for Humans*
edited by Dharmendra K Gupta & Clemens Walther

Vol. 1   *Lithium and Nickel Contamination in Plants and the Environment*
edited by Masahiro Inouhe, Clemens Walther & Dharmendra K Gupta

World Scientific Series on Advances in
Environmental Pollution Management – Volume 3

Valorization of Resources from Urban Mined Materials

Editors

Dharmendra K Gupta
Ministry of Environment, Forest and Climate Change, India

Pankaj Pathak
SRM University AP, India

Dheeraj Mittal
Ministry of Environment, Forest and Climate Change, India

World Scientific

NEW JERSEY · LONDON · SINGAPORE · BEIJING · SHANGHAI · HONG KONG · TAIPEI · CHENNAI

Published by

World Scientific Publishing Co. Pte. Ltd.
5 Toh Tuck Link, Singapore 596224
USA office: 27 Warren Street, Suite 401-402, Hackensack, NJ 07601
UK office: 57 Shelton Street, Covent Garden, London WC2H 9HE

Library of Congress Cataloging-in-Publication Data
Names: Gupta, Dharmendra Kumar (Phytoremediation researcher), editor. |
    Pathak, Pankaj, editor. | Mittal, Dheeraj, editor.
Title: Valorization of resources from urban mined materials / editors, Dharmendra K Gupta,
    Ministry of Environment, Forest and Climate Change, India; Pankaj Pathak, SRM University AP,
    India; Dheeraj Mittal, Ministry of Environment, Forest and Climate Change, India.
Description: Hackensack, NJ : World Scientific, [2025] | Series: World Scientific Series on
    Advances in Environmental Pollution Management ; volume 3 |
    Includes bibliographical references and index.
Identifiers: LCCN 2024033908 | ISBN 9789811297748 (hardcover) |
    ISBN 9789811297755 (ebook for institutions) | ISBN 9789811297762 (ebook for individuals)
Subjects: LCSH: Recycling (Waste, etc.) | Salvage (Waste, etc.) | Mines and mineral resources.
Classification: LCC TD794.5 .V35 2025 | DDC 363.72/82--dc23/eng/20241108
LC record available at https://lccn.loc.gov/2024033908

British Library Cataloguing-in-Publication Data
A catalogue record for this book is available from the British Library.

For any available supplementary material, please visit
https://www.worldscientific.com/worldscibooks/10.1142/13971#t=suppl

Desk Editors: Aanand Jayaraman/Amanda Yun

Typeset by Stallion Press
Email: enquiries@stallionpress.com

Preface

Waste valorization is the practice of transforming waste products or residues into valuable resources through reuse or recycling. This process demonstrates sustainable manufacturing and waste management, which provide economic value to materials that would otherwise be discarded. Generally, it synchronizes the industrial processes, where waste or residues from one process are used as raw materials or energy sources for another.

In the past, industrial processes often regarded waste products as mere byproducts to be disposed of, which frequently led to industrial pollution if not managed properly. However, the landscape began to shift with the rise of increased regulation surrounding residual materials and shifts in socioeconomic attitudes. Starting in the 1990s and continuing into the 2000s, concepts such as sustainable development and the circular economy gained prominence, prompting industries to rethink their approaches to waste. This paradigm shift placed a greater emphasis on recovering resources from waste materials, viewing them as valuable commodities rather than disposable substances. Thus, valorization minimizes waste generation and maximizes resource efficiency. Industrial wastes, due to their consistency and predictability, are prime candidates for valorization efforts. The valorized resource from urban mined materials delves into the crucial topic of efficiently managing waste by transforming it into valuable resources.

Academic research also played a significant role in this shift, with scholars focusing on identifying economic opportunities to reduce the environmental impacts of various industries. For instance, efforts to develop non-timber forest products not only aim to generate economic value but also encourage conservation practices and sustainable utilization. Overall, these developments reflect a growing recognition of the importance of resource efficiency and environmental stewardship in industrial processes.

This book explores various technologies employed for waste valorization, showcasing how materials sourced from diverse origins, such as households, industries, electronic devices, plastics, and packaging materials, can be repurposed through urban mining processes. This comprehensive resource offers insights into innovative approaches that contribute to sustainable waste management practices and the creation of economic value from what was once considered discardable.

The editors, Drs. Dharmendra K. Gupta, Dheeraj Mittal, and Pankaj Pathak, personally acknowledge all the contributing authors for their valuable time, knowledge, and eagerness in bringing this book to its final form.

About the Editors

Dr. Dharmendra Kumar Gupta is Director(S)/Scientist-F at India's Ministry of Environment, Forest and Climate Change, New Delhi. His field of research includes abiotic stress by radionuclides/heavy metals and xenobiotics in plants; antioxidative system in plants, environmental pollution (radionuclides/heavy metals) remediation through plants and microbes (phytoremediation/bioremediation). He has been awarded several international awards by entities such as the Japan Society for the Promotion of Science (JSPS), the Belgian Science Policy Office (BELSPO), JAE-Doc., Spain, TWAS-CNPq, Italy, Royal Society, UK, and MASAV, Israel. Dr. Gupta has published more than 130 internationally peer-reviewed original research/review articles/book chapters and 26 books and served as Book Series Editor for 2 book series from Springer, Switzerland and 1 from World Scientific, Singapore. He has successfully completed 10 multidisciplinary research projects from international and national bodies.

Dr. Dheeraj Mittal graduated from University of Delhi and was awarded a Ph. D. in Plant Molecular Biology in 2013. He joined the Indian Forest Service in the year 2012 and worked as a Forest Officer in the State of Gujarat, India. He has worked in Gir National Park and Wildlife Sanctuary as a protected area manager. He joined the Ministry of Environment, Forest and Climate Change, of the Government of India as Assistant Inspector General of Forest and is working for Project Lion and Elephant (PT&E) and Forest Conservation.

Dr. Pankaj Pathak is an Associate Professor in the Department of Environmental Science and Engineering at SRM University, AP, India. Her research domain includes valorization of resources from solid and hazardous waste, circular economy in metal recovery from e-waste and solar panels. Dr. Pathak is keenly interested in acquiring sustainable secondary resources for a sustainable future. The outcome of her research works has been published in patents, peer-reviewed articles of high-impact international journals, and books.

Contents

Chapter 1

Methodological Proposals for Minimizing Industrial Waste

Edna Regina Amante[*,‡]**, Luiza Helena Meller da Silva**[*]**, and Armando Borges de Castilhos Júnior**[†]

[*]*Programa de Pós-Graduação em Ciência e Tecnologia de Alimentos, Universidade Federal do Pará, Rua Augusto Correa S/N, Guamá, Belém, Pará, 66075-900, Brazil*

[†]*Universidade Federal de Santa Catarina, R. Eng. Agronômico Andrei Cristian Ferreira, s/n - Trindade, Florianópolis - SC, 88040-900, Brazil*

[‡]*eamante.1957@gmail.com*

Abstract

Actions to minimize industrial waste can contribute not only to the reduction of environmental impact but also to the conversion of polluting wastes into new sources of income. The valorization of raw materials and consideration of the different regional cultures of production units represent the achievement of the Sustainable Development Goals. There are countless studies on models for decision-making in industrial waste minimization. However, especially in agriculture and agroindustry, where there is high consumption of water and many raw materials are not yet fully valorized, actions need to be differentiated using accessible models and technologies compatible with the different processes carried out at their respective scales. An example of the diversity of scales and processes is seen in the food industry, which includes a range of small-, median-, and large-sized industries. Environmental care coexists on all the different scales; however, models for waste minimization must be compatible despite these differences. In this work, several historical examples of modeling for waste minimization are presented.

Keywords: Modeling, waste minimization, agroindustry, food industry.

1. Introduction

The need to supply quality food worldwide coexists with the warning that agriculture effectively contributes to climate change mitigation (OECD-FAO Outlook, 2023–2032). In this same sense, consumers concentrate on areas where industrialization is essential, thus posing the challenge of balancing production and industrialization with optimal environmental, social, and economic outcomes. A time when technological development has enabled robotization in different sectors cannot tolerate coexistence with environmental risks.

People develop consumption habits by considering environmental factors, often opting to give up animal-based foods. Currently, there has been growth in consumption of plant-based foods, observing the type of cultivation, processing, packaging, and other particular actions that have created a differentiated market in agriculture and agroindustry. Depending on the practices adopted, it may or may not contribute to reducing environmental impact and requires additional care for the health of consumers who, when adopting a so-called "healthy" diet without proper monitoring, might not incorporate all the essential nutrients in their diets. In this context, consumer behavior is decisive for environmental quality, and the level of importance attributed to individuals as environmental actors in caring for the planet is high (Hoffmann *et al.*, 2024).

The coercive arm of governance in different regions of the world, whether local or regional, has contributed to preventing disasters from becoming even greater by increasing restrictions on the use of pesticides, water consumption, and effluent emissions aimed at protecting the environment. Moreover, these examples should be followed, as in the cases of Norway and Singapore, which took the decision to continue the production and industrialization of fish, maintaining water quality (Bohnes *et al.*, 2022). Another example is "blue production," which involves many actions in different regions of the world, valorizing both production and regional socio-environmental aspects (FAO, 2018).

Despite commendable global actions, the number of initiatives falls short of what is needed to guarantee sustainability for future generations on this planet. New initiatives must be encouraged where operational research can contribute to organizing sectors to minimize waste and control pollution by taking widely applicable actions across several sectors, including agriculture and agroindustry.

Therefore, actions to minimize the environmental impact on the agricultural and agro-industrial sectors must be taken by both developed and developing countries. Considering the contributions of various works and research studies published and put into practice in this context, we present a historical review of the initiatives aimed at creating models for reducing environmental damage, effluent emissions, and waste minimization.

2. Historical Aspects of Modeling for Waste Minimization

Although decision-making in any field is generally multifactorial, it can be particularly difficult in the environmental field for different reasons, especially because of the number and diversity of stakeholders participating in the decision-making.

Interest groups frequently note problems from vastly different points of view, where the potential complexity and perceived importance of environmental decisions give rise to powerful arguments that steer them toward the horizon of a quantitative view (Jennings & Nagarkar, 1996), environmental decision analysis (EDA), or environmental decision-making (EDM), all of which can involve an organized specialty. It has already been established that decision-making regarding environmental problems can be complex and uncertain. However, a deeper knowledge of the different industrial sectors can contribute to increasing sensitivity in decision-making (Haag *et al.*, 2022).

For example, the objectives of wastewater management focus on environmental protection while considering economic, social, and political aspects (Kankasar & Polprasert, 1983). The adoption of optimization techniques such as linear programming, dynamic programming, and comprehensive programming prioritizes studies in the field related to the planning of water sources.

Studies applying operational research have been conducted on a large scale regarding decisions over disposal of urban waste and industrial wastewater (Diop & Maystre, 1989; Kankasar & Polprasert, 1983; Lessard & Beck, 1991; Delgado-Enales *et al.*, 2023; Tomascelli *et al.*, 2024). Those authors demonstrated the application of linear and nonlinear programming in locating water treatment plants and disposing of solid waste from a certain region, minimizing not only the costs of water treatment, including transportation and disposal of solids, but also the adverse impacts on the quality of water.

There are a vast number of publications on wastewater treatment systems due to their importance in environmental engineering, especially regarding the comparison of techniques for the generation of systems. Chang and Liaw (1985) compared two modeling techniques for the generation of systems, analyzing efficient random generation (ERG) and generation and selection (G&S).

Optimization methods can be used in projects and operations of integrated waste management systems, resulting in a decrease in costs and associated difficulties; therefore, the creation of centralized material recovery units has been studied.

Lund *et al.* (1994) stated that linear programming is ideal for preliminary economic studies of wastewater recovery units. On the other hand, Ellis and Tang (1991) stated that linear programming is unsuitable for handling optimization processes because the relationships between different wastewater treatment processes are usually nonlinear.

Additionally, studies on nonlinear programming capable of selecting optimal treatment alternatives through the introduction of the "zero-one" decision variable, thus turning the model into a mix of nonlinear and integral programming, were used in the development of a model to optimize wastewater treatments. This not only considers the minimization of cost but also introduces more subjective concepts for developing a model that includes environmental, social, and cultural considerations, thereby working with dozens of variables.

Various other works are dedicated to the development of models for the centralization of regional liquid and/or solid waste treatment. Voutchkov and Boulous (1993) presented the advantages of allocating treatment of regional wastewater to a single plant.

The resulting problem, which is classified as large-scale due to being part of a regional planning process, can be divided into two stages: selection (screening) and optimization. The first goal is to achieve a reduction in the number of possible alternatives for regionalized treatment, with a consequent reduction in the detailing of information. It refers to a more generic and simplified stage of the process, which is simpler than the optimization stage. This consists of finding the minimum cost at a regional level. It involves linear programming, extreme point ordering, dynamic programming, "branch-and-bound" combinatorial programming, and a combination of linear and dynamic programming. The heuristic method is intended for use in the preliminary stages of pre-optimization and can be used in conjunction with any other optimization process in the second stage of large-scale system analysis.

The procedure involving prior selection was also used by Spriggs and Smith (1996), with the justification that the selection stage provides a valuable alternative for solving environmental problems. Ellis *et al.* (1985) already highlighted the advantages of centralized wastewater treatment units and proposed the development of a stochastic optimization/simulation model for the sequence of treatment of industrial waste liquids. The proposed model contains the following critical attributes:

(1) objective selection of the treatment unit process based on cost efficiency;
(2) variability in the concentration of contaminants in effluents incorporated into the model;
(3) ability to accommodate different effluent flows with specific compositions.

This model is a useful tool for outlining sequences of minimum treatment costs for industrial wastewater. The importance of this model lies in characterizing the concentration of contaminants in the influent using the lognormal density probability function with the influent flow generated by the Monte Carlo technique. Ong and Adams (1990) agree with this trend toward centralization of treatment units, which, due to the complexity involved, has led to the search for mathematical models to solve environmental problems.

Most techniques developed to solve this type of model suffer from one or more of the following problems:

(1) Restrictions on processing capabilities cannot be easily imposed.
(2) Existing capacity cannot be easily incorporated.
(3) Computational resources are greatly needed.

In view of this, most techniques offer a comprehensive and simple alternative for solving the problem of expanding a regional effluent treatment system. They consider not only the capacity and intra- and inter-plant transfers but also the quality of the water received during treatment.

According to Adrian *et al.* (1994), the work of Sheeter and Phelps (1925) was the first to develop a relationship between the amount of organic residue, biochemical oxygen demand (BOD), and dissolved oxygen in rivers, producing the classic "sag" model of dissolved oxygen.

Several other contributions have been added, and in 1991, Yu *et al.* introduced the concept of a river's "memory time," i.e., "the amount of time necessary for a river to forget that it received a certain amount of waste." The same authors, cited by Adrian *et al.* (1994), stated that common one-dimensional analytical solutions often ignore the fact that BOD is always time-dependent, so they apply the superposition method to describe the impact of discrete pulses of BOD applied to a river.

Adrian and collaborators (1994) developed a model for the river's response to the sinusoidal variations in BOD, and they suggested its application for parameter optimization.

One of the applications where wastewater could be reused is in soil treatment. Despite this possibility, an entry and exit flow must be followed for the process to be effective. According to Buchberger and Maidment (1989), the problem of sizing a location for storing wastewater in the ground is, in many aspects, similar to the sizing of a water supply reservoir, and the techniques developed in stochastic analysis applied to water supply reservoirs can be highly useful and thus acceptable for solving the problem of wastewater storage in soil treatment operations.

Considering the importance of urban solid waste processing systems, Ceric and Hlupic (1993) developed a complete simulation study. A conceptual model of the system was introduced using the activity cycle diagram technique, and this conceptual model was used to generate the simulation program.

The authors stated that the disposal of industrial and urban solid waste in landfills culminates in the saturation of these deposits, requiring administrative actions related to placing covering layers and monitoring their effectiveness. Additionally, new areas must be found relatively frequently. At the same time, recycling is an option, but landfills will persist since not everything is recyclable, which became apparent in the 1990s. Therefore, alternatives are found through the circular economy, where companies and communities have the prospect of minimizing the disposal of waste by transforming it into new products and income (Hortono *et al.*, 2023). According to Lambert and Gupta (2004), remanufacturing recreates new products using components from the end of life of old products. Unlike older studies, many recent studies consider the importance of metaheuristics and robotic cells in solving mathematical models for decision analysis of environmental problems (Hortono *et al.*, 2023).

Lund (1990) developed a method that uses linear programming to evaluate and plan the recycling of municipal solid waste. That method provides a minimum-cost recycling plan and maximizes the useful life of landfills. Recycling options minimize the current cost of disposal of solid waste. In that method, the engineer must describe each recycling option in terms of its implementation costs and average efficiency in reducing disposal volume. That author draws attention to the need to update recycling options considering the realities of the period of interest. After developing the model, a computer program is used to solve the problem.

Specific models for certain agricultural and/or industrial activities have been developed. For example, Bouzaher and Offutt (1992) developed a mathematical model to ascertain the feasibility of converting corn harvest residues into alcohol. From a methodological point of view, the authors used three integrated operational research tools: linear programming, Monte Carlo simulation, and constraint exchange programming.

Operational research studies focus mainly on solving problems with a large number of variables, with the help of expert systems and computer programs that speed up the resolution of algorithms. The environmental field has several specific programs within the following categories: interpretation, planning, prevention, diagnostics and repairs, monitoring and control, and instructions and projects (Hushon, 1990).

Ford *et al.* (1993) compared two computer programs, Agricultural Waste Application Rule-based Decision Support (AWARDS) and Ground Water Loading Effects of Agricultural Management System (GLEAMS), which are different in complexity since AWARDS is a program that allows farmers to plan and manage all their properties and the application of residues, while in GLEAMS, the most important aspects are related to the mass of nitrogen and/or phosphorus in kg/ha and their movement in the soil toward the roots of plants. The AWARDS program requires 1,741 parameters for each field to be evaluated, whereas GLEAMS requires only 248 parameters for the simulation. The authors conclude that although the two programs have different levels of technology, they can be used in combination to assist rural producers (farmers) in their agricultural work in harmony with the environment.

Many efforts have been made to resolve the problems arising from multidisciplinarity and differences in scale and applications across multiple sectors. Recently, Maghami *et al.* (2024) presented a flexible and scalable method for describing hydrologic and environmental models with metadata using a machine-readable JSON schema. The extensible model is also expected to better support the reproducibility of model-related research, which is an essential issue across all scientific disciplines, by facilitating the sharing, discovery, and reuse of data and models built by others.

The different production chains in the food industry that contribute to the manufacture of products until the end of their useful lives have contributed to the consumption of water and energy and the generation of waste, thus creating great demand for solutions aimed at achieving sustainable development (UNEP, 2017). Therefore, this is a sector that involves a great diversity of scales, ranging from small (family businesses, with few employees) to large (big or multinational companies, with many employees), and on all the different scales, the demand for environmental solutions is notable. Therefore, the great diversity of production scales in the food industry requires a solution model that is compatible with the realities of this sector.

Consequently, regardless of the production scale, all waste minimization projects must be accompanied by environmental, economic, social, and technical feasibility analyses that meet the United Nations' 17 Sustainable Development Goals (UN SDGs). However, another reality in this sector is the need for a language that is accessible to the production sector. Thus, Rezzadori *et al.* (2012) and Oliveira *et al.* (2013) presented examples of waste valorization in a language that is accessible to professionals with different backgrounds in the food sector.

3. Model for Waste Minimization Taking into Account the Concepts of Socio-Environmental Actors

The environmental management model for the food industry is exemplified and proposed based on the structure for using systems and procedures for environmental management, as presented in Figure 1.

Systems and procedures for waste management, including their application as by-products or in new products, can achieve either high or low efficiency levels. Such levels can be proven by the large number of scientific studies, which are disproportionately outweighed by the benefits for global environmental issues.

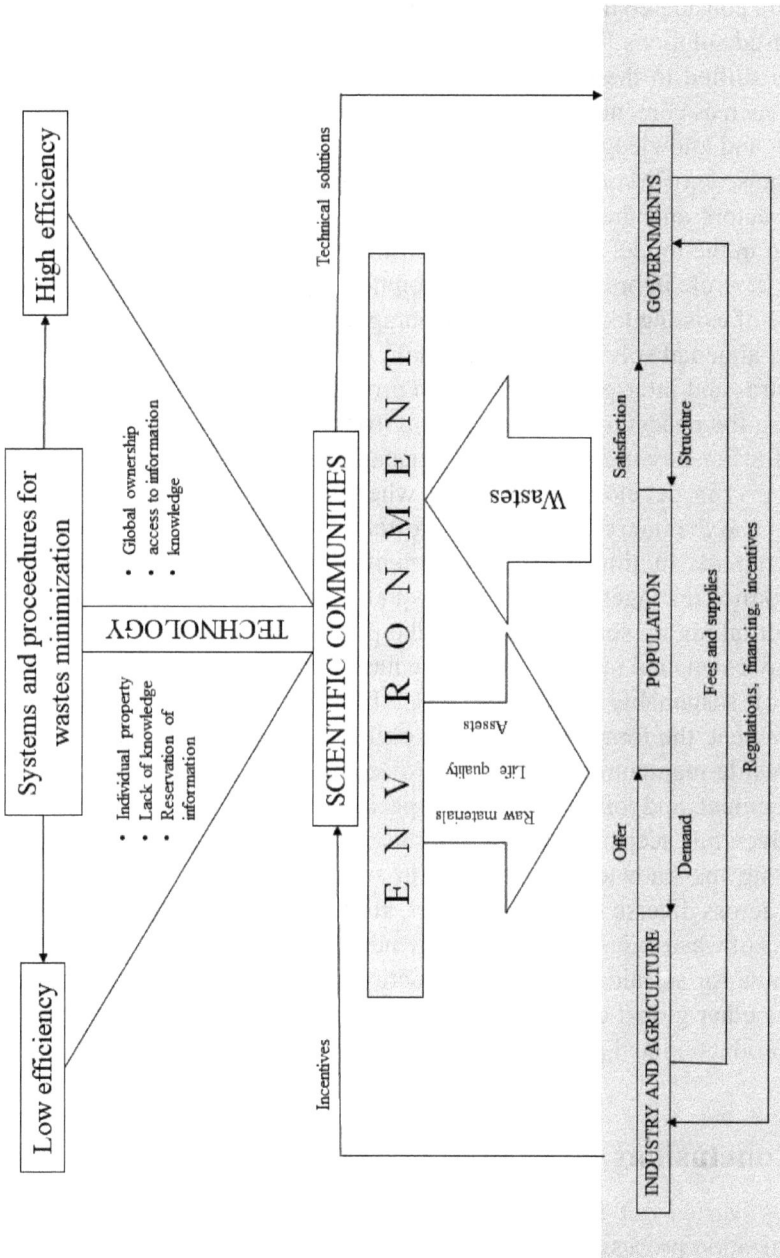

Figure 1. Scenario of the effects of the actions of different social actors on efficiency in environmental management.

Individual ownership, information reservation, and a lack of knowledge are considered the main factors that contribute to the wide prevalence of unused solutions for the benefit of the global environment. Efficiency can be shifted to the right (Figure 1) through behaviors contrary to this trend, such as contemplating global ownership and greater access to information and knowledge.

The scientific community serves as a link between various groups of social actors and the available procedures or systems, which acts as a vehicle in the model presented. Industries, governments, and populations play a key role in providing the environment with due protection by making use of existing techniques or encouraging the conduct of new research studies aimed at solving environmental problems.

Intra- and inter-group integration problems may have contributed to framing the necessary ideal global environmental objectives. The problem of efficiency in environmental management tends to shift to the left (Figure 1) in developing countries, where it is common for economic factors and the interests of groups to often outweigh the real needs of the environment. In this way, environmental management gives way to "economic" management, where the quality of life on the planet persists without taking up space in competitive processes.

Environmental management procedures aimed at achieving the objectives of sustainable development (UNEP, 2017) demand, as a basic requirement, the formation of a work philosophy focused on the environment while maintaining the nuances of competitiveness, production and development, and long-term work, especially in developing countries.

Given this scenario, insertion in the production sector is essential for acquiring the knowledge necessary to suggest measures that minimize waste across diverse sectors. However, all the actors can contribute to the success of waste minimization management, which justifies the success of programs for sustainable waste production in Norway and Singapore, as well as other global examples, such as the movement toward sustainable food production and processing.

4. Conclusions

Studies carried out to develop mathematical models to support waste minimization processes are increasing and must meet the expectations of demand in environmental projects.

Decision-making to minimize waste in the food sector coexists with its great diversity of scale and the need for sectorized actions, aiming to recognize the chemical properties of solid and liquid waste, reduce effluent treatment operations, and increase the full recovery of raw materials. The food industry is an important example of this diversity, where proposals can be presented and decisions about valuing raw materials and minimizing waste can become reality through effective organization of the sector and environmental actors, increasing the permeability of information and initiatives that address the problem in a sectorized way, regardless of the production scale.

Encouraging the creation of knowledge bases and the permeation of information about various industrial sectors presents a great opportunity for waste management, recognizing the need for innovation, so that waste treatment units are minimized and raw materials are valued within or between sectors.

References

Adrian, D. D., Yu, F. X., & Barbe, D. (1994). Water quality modeling for a sinusoidally varying waste discharge concentration. *Water Research, 28*, 1167–1174.

Bohnes, F. A., Hauschild, M. Z., Schlundt, J., Nielsen, M., & Laurent, A. (2022). Environmental sustainability of future aquaculture production: Analysis of Singaporean and Norwegian policies. *Aquaculture, 549*, 737717.

Bouzaher, A. & Offutt, S. (1992). A stochastic linear programming model for corn residue production. *Journal of the Operational Research Society, 43*, 843–857.

Ceric, V. & Hlupic, V. (1993). Modeling a solid-waste processing system by discrete event simulation. *Journal of the Operational Research Society, 44*, 107–114.

Chang, S.-Y. & Liaw, S.-L. (1985). Generating designs for wastewater systems. *Journal of Environmental Engineering, 111*, 665–679.

Delgado-Enales, I, Der, S. J., & Molina-Costa, P. (2023). A framework to improve urban accessibility and environmental conditions in age-friendly cities using graph modeling and multi-objective optimization. *Computers, Environment and Urban Systems, 102*, 101966.

Diop, O. & Maystre, L. Y. (1989). Méthodologie Systémique multicritère appliquée à la gestion des déchets solids urbains de Dakar (Senegal). *Tech Sci Met - L'eau, 84*, 187–191.

Ellis, J. H., McBean, E. A., & Farquhar, G. J. (1985). Stochastic optimization/ simulation of centralized liquid industrial waste treatment. *Journal of Environmental Engineering, 111*, 804.

Ellis, K. V. & Tang, S. L. (1991). Wastewater treatment optimization model for developing world. I: Model development. *Journal of Environmental Engineering, 117*, 501–517.

Food and Agriculture Organization (FAO) (2018). The FAO blue growth initiative: Strategy for the development of fisheries and aquaculture in eastern Africa. Available at: https://www.fao.org/3/i8512en/I8512EN.pdf (Accessed 3 February 2024).

Ford, D. A., Kruzic, A. P., & Doneker, R. L. (1993). Using GLEAMS to evaluate the agricultural waste application rule-based decision support (AWARDS) computer program. *Water Science and Technology, 28*, 625–634.

Haag, F., Miñarro, S., & Chennu, A. (2022). Which predictive uncertainty to resolve? Value of information sensitivity analysis for environmental decision models. *Environmental Modelling & Software, 158*, 105552.

Hoffmann, T., Ye, M., Zino, L., Cao, M., Rauws, W., & Bolderdijk, J. W. (2024). Overcoming inaction: An agent-based modelling study of social interventions that promote systematic pro-environmental change. *Journal of Environmental Psychology, 94*, 102221.

Hortono, N., Hamírez, F. J., & Pham, D. T. (2023). A multiobjective decision-making approach for modelling and planning economically and environmentally sustainable robotic disassembly for remanufacturing. *Computers & Industrial Engineering, 184*, 109535.

Hushon, J. M. (1990). Expert systems for environmental problems. *Environmental Science and Technology, 21*, 838–841.

Jennings, A. A. & Nagarkar, P. A. (1996). Automating probabilistic environmental decision analysis. *Environmental Software, 10*, 251–262.

Kankasar, B. R. & Polprasert, C. (1983). Integrated wastewater management. *Journal of Environmental Engineering, 109*, 619–630.

Keykhosravi, M., Dehyouri, S., & Mirdamadi, S. M. (2023). Modeling the environmental performance by focusing on environmental behavior rural farmers. *Environmental and Sustainability Indicators, 20*, 100309.

Lambert, A. F. & Gupta, S. M (2004). *Disassembly Modeling for Assembly, Maintenance, Reuse and Recycling.* CRC Press, Boca Raton, Florida, USA.

Lessard, P. & Beck, M. B. (1991). Dynamic modeling of wastewater treatment processes. *Environmental Science & Technology, 25*, 30–39.

Lund, J. R. (1990). Least-cost scheduling of solid waste recycling. *Journal of Environmental Engineering, 116*, 182–197.

Lund, J. R., Tchobanoglous, G., Anex, R. P., & Lawer, R. A. (1994). Linear programming for analysis of material recovery facilities. *Journal of Environmental Engineering, 120*, 1082–1094.

Maghami, I., Morsy, M. M., Sadler, J. M., Horsburgh, J. S., Dash, P. K., Choi, Y., Chen, K., Seul, M., Black, S., Tarboton, D. G., & Goodall, J. L. (2024). An extensible schema for capturing environmental model metadata:

Implementation in the HydroShare online data repositor. *Environmental Modelling & Software, 172*, 105895.

OECD-FAO-2023-2032 Outlook (2023–2032). https://www.oecd-ilibrary.org/docserver/08801ab7-en.pdf?expires=1706810147&id=id&accname=guest&checksum=81F90EBFDF75529583435DB89EEF2701 (Consulted 1 February 2024).

Oliveira, D. A., Benelli, P., & Amante, E. R. (2013) A literature review on adding value to solid residues: Egg shells. *Journal of Cleaner Production, 46*, 42–47.

Ong, S. L. & Adams, B. J. (1990). Capacity expansion for regional wastewater systems. *Journal of Environmental Engineering, 116*, 542–560.

Rezzadori, K., Benedeti, S., & Amante, E. R. (2012) Proposals for the residues recovery: Orange waste as raw material for new products. *Food Bioproducts Processing, 90*, 606–614.

Spriggs, H. D. & Smith, W. R. (1996). Design for pollution control: Screening alternative technologies. *Environmental Progress, 15*, 69–72.

Tamascelli, N., Dal Pozzo, A., Scarponi, G. E., Paltrinieri, N., & Cozzani, V. (2024). Assessment of safety barrier performance in environmentally critical facilities: Bridging conventional risk assessment techniques with data-driven modelling. *Process Safety and Environmental Protection, 131*, 294–311.

UNEP (2017). *United Nations Environmental Programme*. Technical Report. United Nations. https://www.unep.org/es.

Voutchkov, N. S. & Boulos, P. F. (1993). Heuristic screening methodology for regional wastewater-treatment planning. *Journal of Environmental Engineering, 119*, 603–644.

Chapter 2

Metals Recovery from Electrical and Electronic Waste

Aswetha Iyer*, Krishnanjana S. Nambiar*, Dharmendra K. Gupta†, and S. Murugan*,‡

*Department of Biotechnology, Karunya Institute of Technology and Sciences (Deemed to be University), Coimbatore 641 114, India
†Ministry of Environment, Forest and Climate Change, Indira Paryavaran Bhawan, Jorbagh Road, Aliganj, New Delhi 110003, India
‡micromurugans@gmail.com

Abstract

This chapter delves into the critical realm of metal recovery from electronic waste (E-waste), addressing the pivotal role it plays in alleviating both resource depletion and environmental pollution. With an intricate exploration of diverse recovery methodologies, such as hydrometallurgy, pyrometallurgy, and emerging bioleaching, the chapter elucidates their environmental considerations and strengths. Essential preliminary processes, including proper E-waste collection, dismantling, and sorting, are highlighted as prerequisites for effective metal recovery. Technological advancements and their impact on enhancing efficiency and yield are underscored, accompanied by an analysis of environmental implications, emissions, and residue management. The economic dimensions of E-waste metal recovery, encompassing market dynamics and potential revenue streams, are comprehensively examined. Emphasizing a circular economy, the chapter advocates for a holistic approach that amalgamates technological innovations, eco-friendly practices, and economic viability to maximize metal recovery and minimize the environmental impact of E-waste. Keywords encompass resource depletion, the circular economy, waste management, clean technology, and the imperative need for sustainable practices. This comprehensive exploration aims to guide stakeholders toward responsible and effective strategies in the intricate landscape of E-waste metal recovery.

Keywords: Metal recovery, electronic waste, hydrometallurgy, pyrometallurgy, bioleaching, circular economy.

1. Introduction

Metals are inorganic substances made up of a combination of metallic and non-metallic elements with properties such as hardness, luster, malleability, fusibility, and ductility and also characterized by electrical and thermal conductivity (Minay & Boccaccini, 2005). These properties make them essentially important to carrying out day-to-day activities and thus hard to replace in our lives. Metals are used in various industries to produce jewelry, electrical wires, utensils, etc., with manufacturing industries utilizing 31% of metals, mostly iron and steel. However, long-term exposure to these elements can lead to metal poisoning, which results in the development of neurological disorders (Chen *et al.*, 2016).

The metals copper (Cu), lithium (Li), tin (Sn), silver (Ag), gold (Au), nickel (Ni), and aluminum (Al) are the most commonly used in electrical and electronic devices, such as computers, capacitors, and refrigerators. Electronic waste (E-waste) from these devices is categorized as non-biodegradable and is harmful to nature. E-waste components, such as cadmium (Cd), lead (Pb), lead oxide (PbO), antimony (Sb), nickel (Ni), and mercury (Hg), are released into the surroundings, polluting the ecosystems (Dissanayake, 2014). Exposure to heat causes the release of carcinogenic gases and dust particles as dioxins that damage the respiratory system, and the percolation of these toxic elements into the groundwater and soil affects the food chain and productivity of land, respectively. The conventional methods for the disposal of E-waste are landfilling, acid bath, incineration, recycling, and the reuse of the elements (Monika & Kishore, 2010). With incineration and landfilling, toxic components are easily released into the environment; on the other hand, acid baths, which are used to break down all the elements, can lead to burns or other health issues. Hence, there is a need to improve the strategies for disposing of E-waste. One of the effective methods is recycling E-waste, as its primary goal is to bring down the pollution caused by E-waste, lessen the consumption of resources, and prevent the leaching of elements as much as possible (Jadhao *et al.*, 2022). In addition to this, technological advances have paved the way for strategies that aim to extract or retrieve the metals for use in industries through a process called "recovery." This method helps in the recovery of precious metals, such as Au, Ag, Pt, Cu, Al, and Ni, from E-waste using hydrometallurgical (HM) or pyrometallurgical (PM) techniques (Manikandan *et al.*, 2023). Recently, efforts have been made to use microorganisms for the recovery of metals through the process of biohydrometallurgy due to the drawbacks of HM and PM (Desmarais *et al.*, 2020; Free, 2014; Manikandan *et al.*, 2023).

1.1 *Types of E-waste*

The seven types of E-waste are illustrated in Figure 1.

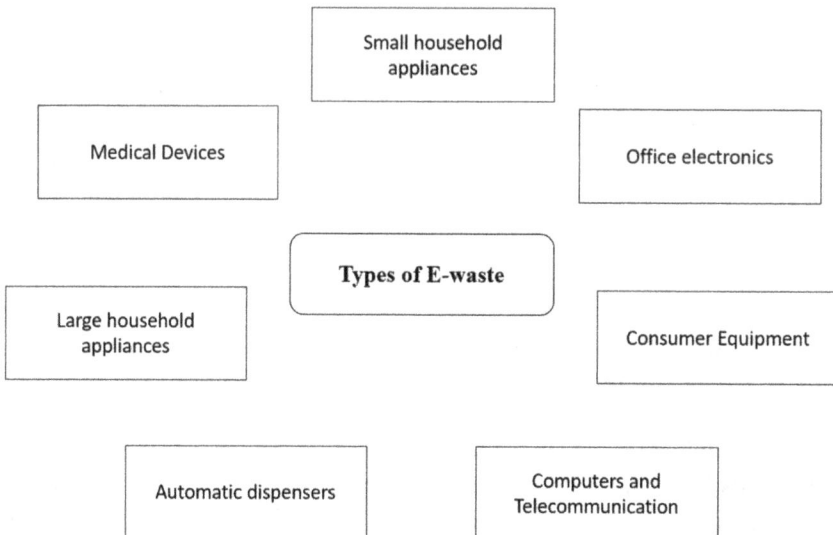

Figure 1. Types of E-waste.

Table 1.   Sources of E-waste and the metals used in them.

Source of E-waste	Metals present	References
Television	Cu, Quartz, Fe, Au, Pt, Ag	Ari (2016)
Computer	Pd, Cu, Ni, Ta, Si, Au, Al, Zn, Fe	Abbasi (2023)
Home appliances	Cu, Au, Ag, Al, Steel, Brass, Mg	Cleaver *et al.* (2014)
Children's toys	Ar, Sb, Cd, Pb, Cr, Hg, Cu, Co, Ti, Zn, SI, Fe	Ahmed *et al.* (2021)
Solar panels	Si, In, Ga, Se, Cd, Te, Cu, Ni, Zn, Ta	Chinmay (2022)
LED bulbs	Mn, La, Y, BaO/AlO	Harris (2010)
Agricultural Equipment	Steel	Reed (2021)
Vending machine	Galvanized steel	Vending Machine (n.d.)
Medical devices	Stainless steel, Cu, Ti, Co, Al, Mg, Au, Pt, Au, Ir, Ta	Ulbrich Stainless Steels & Special Metals, Inc. (2021)
Space shuttles and devices	Al, Au, Ti, Ni	Shulman (2017)
Automobiles	Steel, Fe, Al, Mg, Pb, Cu, Ti	Briffa (2020)

1.2 *Sources and distribution of E-waste*

The main sources of E-waste include televisions, computers, mobile accessories, home appliances, air conditioners, and children's toys.

The main metallic compounds used in each of these are listed in Table 1.

2. Importance of Recovering Metals from E-Waste

Growing concerns about the percolation of metals into aquatic and terrestrial habitats eventually led to the rise of techniques that help in reducing pollution and mitigating disturbance to the ecosystem and the food chain. The adverse leaching into the environment can interfere with an organism's biological and metabolic functions (Tchounwou *et al.*, 2012). Reusing the recovered metals is another strategy for the sustainable use of metals in the future (Ali *et al.*, 2019). The adopted methodologies for the recovery of metals do not change their intensive properties but provide benefits by improving the quality of air and lowering the use of energy and resources, in addition to reducing the wastage of tons of metal reservoirs, thus focusing on developing an approach for circular economy (Shamsuddin, 1986; Sree Lakshmi *et al.*, 2021).

2.1 *Corollary impacts of metals on the environment and the estimation of pollution caused by metals*

The metals from E-waste are released into the environment due to the processes of burning, shredding, and dismantling, causing the redeposition of these elements back into the soil due to their size and weight and leading to the accumulation of toxic compounds in the ecological pyramids of the food web (Tchounwou *et al.*, 2012). Water pollution follows due to the leaching of metals into the water resources, where these elements undergo processes such as acidification and toxification, negatively impacting the lives of aquatic organisms. Methods such as landfilling, acid baths, and incineration have contributed more to the release of toxic metals, including Hg, Cd, Be, and Pb, directly into the environment (Jain *et al.*, 2023).

Metals in trace amounts are important to every organism, but exceeding these optimum requirements causes chaos in any biological system. In plants with high metal uptake, they can reduce metabolic capabilities, thus enhancing the generation of reactive oxygen species (ROS). This inhibits plant growth and alters the composition of its microbiome (Jain *et al.*, 2023; Kiran & Sharma, 2022; Singh & Kalamdhad, 2011; Gupta *et al.*, 2015). At the same time, in animals, heavy metals compete with other substrates for the active sites of enzymes, thereby inhibiting the function of the cells, which subsequently affects their internal organs (Govind & Madhuri, 2014; Singh & Kalamdhad, 2011). Aquatic life is affected by DNA damage, the generation of ROS, and changes in pH, causing disruptions to their nervous system and eventually leading to death (Garai *et al.* 2021). In human beings, the accumulation of metals can cause the malfunctioning of organs and create a competitive environment for enzyme sites (Singh *et al.*, 2011).

According to a report from the World Health Organization (WHO), around 53.6 million tons of E-waste were estimated to be produced in 2019. At the same time, records estimate the production of about 50 million tons of E-waste every year, out of which only 17% are stored and

Table 2.   List of metals, methods, and industries.

Metals	Recovery methods	Industries	References
Gold	Macarthur forest cyanide process (hydrometallurgy) and selective polymeric film	The extracted gold is used for the production of new models of jewels, in dental clinics, and in electronics by melting the scrap gold	Manikandan et al. (2023), Gasdia (2023), Karelia (2021), and Gulliani et al. (2023)
Silver	Melting in plasma reactors (pyrometallurgy), liquid membrane technology, and other hydrometallurgical processes	Reused in jewelry and electronics	Čarnogurská et al. (2018), Kahar et al. (2023), and Manikandan et al. (2023)
Platinum	Acid/base-based metal leaching (HM) and PM methods such as smelting, vaporization, and sintering.	Used in glass manufacturing industries, quantitatively in the chemical or pharmaceutical industry for its medical capabilities	Cayumil et al. (2016), Ding et al. (2019), Islam et al. (2020), Manikandan et al. (2023), and Singh (2021)
Aluminum	Using High-Force® eddy-current separators and three-layer electrorefining process apparatus	Mostly used in the manufacturing of sheet aluminum and home accessories, in the automobile industry, and in mending electrical wires	David (2021), Manikandan et al. (2023), and Zhang et al. (1998)
Nickel	By fluidized-bed non-seeded granulation process and by electrodeposition	For the quantitative production of coins, steel, musical instruments, airships, and other tools and in heat and electricity generation	Coman et al. (2013), Manikandan et al. (2023), and Robotin et al. (2013)

Zinc	By electrodeposition method	Due to its corrosion protection properties, it is used for automotive and construction activities, in the production of batteries, and in the pharmaceutical and food industries	Manikandan et al. (2023) and Sivakumar et al. (2022)
Copper	Through the chemical leaching process (HM), electrowinning, solvent extraction, and precipitation.	These are recycled for making plumbing tools, roofing sheets, and heat exchangers	A Copper Alliance Member (2015), Hayati et al. (2023), and Manikandan et al. (2023)
Tin	Through a recently developed method called chemical vapor transport process	For fabricating home organizing accessories such as utensils or pen holders, vases, or even garlands	Jha et al. (2012) and Manikandan et al. (2023)
Lead	Direct leaching-electrowinning in calcium chloride solution (HM) and Parkes zinc-desilvering process (PM)	The recovered lead can be reused to make batteries and for roofing, X-ray shielding, and soundproofing	Manikandan et al. (2023) and Xing et al. (2019)
Selenium	Recovered through the process of leaching for selective separation of metals	Used in photocells and solar cells	Manikandan et al. (2023) and Sarangi et al. (2023)
Tellurium	Recovered through HM practices and other techniques, such as adsorption, or bioreduction	Mostly used as an alloy in electrical appliances	Chen et al. (2023) and Manikandan et al. (2023)

(Continued)

Table 2.   (*Continued*)

Metals	Recovery methods	Industries	References
Iron	Using HM methods of adsorption, ion exchange, etc.	Recovered iron can be used for the construction of buildings, pipes, valves, and so on	Ashiq *et al.* (2019) and Manikandan *et al.* (2023)
Cobalt	Recovered through the efficient process of membrane solvent extraction.	Used for the production of motors, required for electronics, and the manufacture of circuit boards	Ciolea *et al.* (2022), Kostanyan *et al.* (2023), and Manikandan *et al.* (2023)
Mercury	Recovered by means of precipitation with aluminum or by electrolysis.	Used in thermometers, barometers, and electrical switches	Britannica (2016) and Manikandan *et al.* (2023)
Antimony	Using HM and electrochemical techniques	To manufacture heat-resistant materials in plastics and electronics	Barragan *et al.* (2020) and Manikandan *et al.* (2023)

recycled, according to the WHO. The rest are believed to be either deposited in landfills or incinerated. The main contributors of E-waste are China, which produces 10.1 million tons, followed by the US, which produces 6.9 million tons, and India, which produces 3.2 million tons, together contributing around 40% of total E-waste.

2.2 Retrievable metals from E-waste

The deficit of metals from their primary sources has made it challenging for the industries to attain their resource requirements, hence forcing them to develop strategies to recover metals, such as Au, Ag, Pt, Cu, Al, Ni, Zi, Tin, Pb, Se, Te, Li, Fe, Co, Hg, Be, In, Cd, As, and Sb, from the E-waste to efficiently reuse them for the development of their products (Manikandan *et al.*, 2023).

The type of metal, method of recovery, and industries reusing these metals are listed in Table 2.

3. Strategic Methods for the Recovery of Metals

The reprocessing of metal scraps can be done through three main steps: remelting, recasting, and redrawing (Britannica, 2023). Metal scraps can be processed into metals for reuse in different industries, and hence sustainable methods are available for the recovery of metals from E-waste to maximize their utilization. Figure 2 shows different types of metal recovery methods that are discussed below.

3.1. Hydrometallurgy

The leaching agent commonly used for the recovery of metals from scraps and ores is dilute sulfuric acid (Dil. H_2SO_4). A strong or watery solution of this acid is used for the selective dissolution and precipitation of metals (Qi, 2018). The precipitated metals are then purified through the sequential steps of extraction, adsorption, and ion exchange to concentrate the metal of interest (Ashiq *et al.*, 2019).

The processes involved in three major steps are as follows:

(a) *Leaching*

It is the process of extracting or losing a material when in contact with an aqueous solution. One of the most commonly extracted metals is

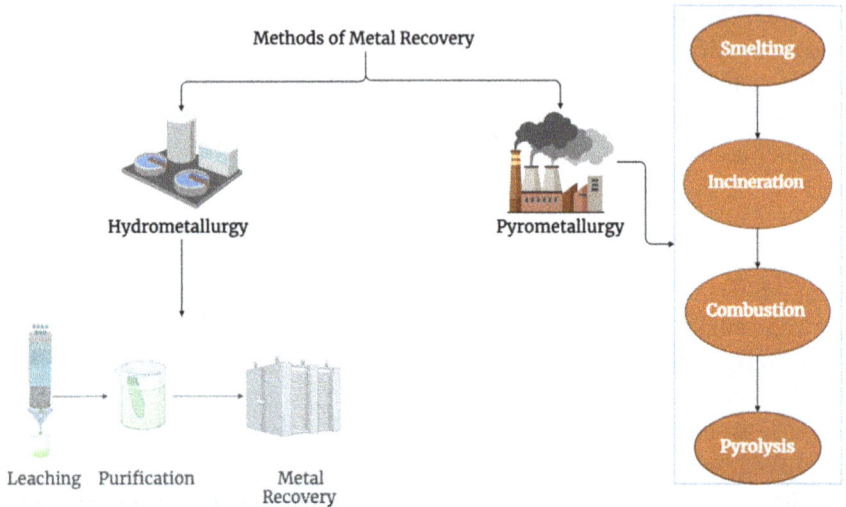

Figure 2.   Types of metal recovery methods.

gold, using a sodium cyanide solution (Yahya *et al.*, 2020). This method of cyanidation helps in the selective extraction of gold from the ore, as the cyanide dissolves the gold from the solution. The extracted gold remains as $Au(CN)_2$ and hence can be recovered from cyanide through the use of magnetic activated carbons (MACs). These MACs consist of the properties of adsorption, enabling the adsorption of gold into them, and they are also reusable (Xia *et al.*, 2021).

(b) *Solution purification (SP)*

The cementation of the compound is the process of agglomeration of compounds or particles, thus concentrating the metals in the leach liquor with the addition of zinc dust along with antimony oxide or arsenic oxide under a pH of 4 (Free & Moats, 2014).

SP ensures the removal of impurities from the solution. Impurities can become one of the major detrimental factors in electrowinning, which is the process of recovering metals such as gold, silver, and copper from an electrolytic solution (Free & Moats, 2014).

(c) *Metal recovery*

The extracted and purified metals can be recovered through the processes of electrolysis, using the technique of gaseous reduction of metals, preferably with H^+ (Bhandari & Dhawan, 2023) or using some means of precipitation with agents such as Ca $(OH)_2$ and NaOH (Tripathee *et al.*, 2019).

3.2 *Pyrometallurgy*

Pyrometallurgy is a method developed for the purification of metals extracted under high temperatures (Singh & Schwan, 2011) and is predominantly used for the treatment of E-waste. The processes involved in PM are smelting, incineration, combustion, and pyrolysis (Murugappan & Karthikeyan, 2021). With the application of heat, metal ores are transformed into oxides. On further processing, metals are released as gases (Britannica, 2016).

4. Guidelines for Metal Disposal and Metal Recovery

4.1 *Transportation*

The movement of waste from production or recycling to a treatment, storage, and disposal facility must adhere to the rules of the Hazardous and Other Wastes (Management and Transboundary Movement) Rules, 2016. In compliance with E-waste management regulations, stakeholders in the E-waste management chain, such as producers and recyclers, are required to store E-waste for a maximum of 180 days, with an extension possibility of up to 365 days for research and development purposes. Thorough records of collection, sale, transfer, and storage must be maintained and available for inspection. Storage practices should the prevent breakage of end-of-life products and ensure worker safety. Specific precautions are needed to avoid damage to refrigerators, air-conditioners, cathode ray tubes, fluorescent lamps, and equipment containing asbestos or ceramic fibers, aiming to prevent environmental releases of hazardous substances. Adherence to these scientific guidelines promotes responsible E-waste management and environmental sustainability.

4.2 *Storage*

Loading, transporting, unloading, and storing of E-waste and end-of-life items should be done in such a way that their ultimate purpose, such as reuse after refurbishment, recycling, or recovery, is unaffected. The storage room should have a fire prevention system installed. Collection centers, which are developed by manufacturers, refurbishers, dismantlers, and recyclers, play an important part in E-waste channelization. Authorized under Extended Producer Reusability (EPR), these facilities collect and store E-waste on behalf of stakeholders before transferring it to authorized dismantlers or recyclers. Only EPR-authorized centers can function, including those founded by dismantlers, recyclers, or refurbishers who have signed agreements.

4.3 *Recycling*

Recyclers, integral to E-waste management, are mandated to build facilities for secure data destruction in end-of-life products like hard disks, phones, and mobiles using methods such as shredding or grinding. Their functions include dismantling and comprehensive recovery operations with no strict operational constraints, provided they possess the necessary facilities. Occupational safety measures necessitate the use of proper protective equipment for workers, ensuring their well-being. Under the rules, anyone involved in recycling waste electrical and electronic equipment is recognized as a recycler. They can establish collection centers without a separate authorization, streamlining operations. Recyclers can source raw materials, such as waste electronic assemblies or components, from various channels, including producers, Producer Responsibility Organization (PRO), E-waste exchanges, dismantlers, and consumers. This flexible approach reflects the interconnected nature of stakeholders in the dynamic management of E-waste.

The revenues generated under environmental compensation will be stored in a different escrow account by the Central Pollution Control Board and shall be utilized for: the recovery and sorting or end-of-life disposal of uncollected, historical, orphaned E-waste and non-recycled or non-end-of-life disposal of E-waste on which the environment compensation is levied; research and development; incentivizing recyclers; providing financial assistance to local bodies for managing waste management projects; and other purposes as decided by the committee.

5. Recent Advancements in Recovering Metals

The conventional methods are found to be time-consuming, high-energy-demanding techniques, for which new alternative methods have been devised that are more efficient in retrieving the metals from the ores or E-waste (Yu *et al.*, 2020):

(a) *Bioleaching (BL)*

BL, also known as biomining or biohydrometallurgy, is a process in which the interactions between microbes and minerals enable the extraction of metals from E-waste. The microorganisms mostly involved in the leaching of metals are iron- and sulfur-oxidizing bacteria, cyanogenic bacteria, and fungi (Adetunji *et al.*, 2023).

Direct bioleaching is the process of direct involvement of microorganisms that release leaching agents capable of separating metals from the rest of the materials or particles (Valix, 2017). The most commonly used microorganisms are *Acidthiobacillus ferrooxidans, A. caldus, A. thiooxidans, and A. ferrivorans* (Jerez, 2009).

Indirect bioleaching is the process in which the microbial agents are not in contact with the minerals but depend on the cell count for the separation of metals from minerals (Vera *et al.*, 2022).

(b) *Bioelectrochemical*

In the reduction or oxidation of metals using microorganisms, an emerging method called the microbial electrolysis cell (MEC) is capable of producing hydrogen from organic matter with the release of electrons through metabolic activities, resulting in the generation of current with the reduction of electron acceptors. These cells are used for heavy-duty transportation (Dikshit *et al.*, 2023; Evren Tugtaş & Çalli, 2018; Pant *et al.*, 2012).

6. Future Prospects of Metal Recovery

Regardless of the properties of metals, there is a need to develop sustainable methods that can help in the recovery of metals from spent or scrap metallic waste. The emerging techniques are promising for the effective recovery of metals that can be reused. Recent studies on the involvement of microorganisms have gained a lot of attention due to their lower production costs.

7. Conclusion

This comprehensive exploration of metals in E-waste highlights their indispensable role in daily life, with diverse applications in various industries. However, prolonged exposure to these metals can lead to detrimental health effects and environmental pollution. E-waste components, when improperly disposed of, release hazardous substances into the ecosystem, impacting soil, water, and air quality. To address these challenges, recycling E-waste emerges as a crucial strategy, aiming to reduce pollution, conserve resources, and prevent the leaching of harmful elements. Advanced technologies, such as hydrometallurgy, pyrometallurgy, and biohydrometallurgy, offer promising avenues for the efficient recovery of metals from E-waste, promoting environmental sustainability and the circular economy. Furthermore, the documented sources, distribution, and types of E-waste, along with their critical impacts on ecosystems, underscore the urgency of responsible metal recovery practices. As we strive toward a sustainable future, the integration of innovative methods and the adoption of circular economy principles in metal recovery processes will play a pivotal role in minimizing environmental pollution and maximizing the reuse of valuable resources.

References

A Copper Alliance Member (2015). Recycling of copper. Copper Development Association Inc., 2 April.

Abbasi, I. (2023). What materials are used to make computer chips? *AZoM*.

Adetunji, A. I., Oberholster, P. J., & Erasmus, M. (2023). Bioleaching of metals from E-waste using microorganisms: A review. *Minerals, 13*(6), 828. https://doi.org/10.3390/min13060828.

Ahmed, H. O., Attaelmanan, A. G., AlShaer, F. I., & Abdallah, E. M. (2021). Determination of metals in children's plastic toys using X-ray florescence spectroscopy. *Environmental Science and Pollution Research, 28*(32), 43970–43984. https://doi.org/10.1007/s11356-021-13838-1.

Ali, H., Khan, E., & Ilahi, I. (2019). Environmental chemistry and ecotoxicology of hazardous heavy metals: Environmental persistence, toxicity, and bioaccumulation. *Journal of Chemistry, 2019*(1), 6730305. https://doi.org/10.1155/2019/6730305.

Ari, V. (2016). A review of technology of metal recovery from electronic waste. In *E-Waste in Transition - From Pollution to Resource*. IntechOpen, Croatia. https://doi.org/10.5772/61569.

Ashiq, A., Kulkarni, J., & Vithanage, M. (2019). Hydrometallurgical recovery of metals from E-waste. In *Electronic Waste Management and Treatment Technology*, pp. 225–246. https://doi.org/10.1016/B978-0-12-816190-6.00010-8. Elsevier.

Barragan, J. A., Ponce de León, C., Alemán Castro, J. R., Peregrina-Lucano, A., Gómez-Zamudio, F., & Larios-Durán, E. R. (2020). Copper and antimony recovery from electronic waste by hydrometallurgical and electrochemical techniques. *ACS Omega, 5*(21), 12355–12363. https://doi.org/10.1021/acsomega.0c01100.

Bhandari, G. S. & Dhawan, N. (2023). Gaseous reduction of NMC-type cathode materials using hydrogen for metal recovery. *Process Safety and Environmental Protection, 172*, 523–534. https://doi.org/10.1016/J.PSEP.2023.02.053.

Briffa, J. (2020). Heavy metal pollution in the environment and their toxicological effects on humans. *Heliyon*, September. https://doi.org/10.1016/j.heliyon.2020.e04691.

Britannica (The Editors of Encyclopaedia) (2016). Pyrometallurgy. *Encyclopedia Britannica.*

Britannica (The Editors of Encyclopaedia) (2023). Recycling. *Encyclopedia Britannica.*

Čarnogurská, M., Příhoda, M., Lázár, M., & Kurilla, P. (2018). Pyrometallurgical treatment of silver-containing catalysts. *Materiali in Tehnologije, 52*(2), 133–138. https://doi.org/10.17222/mit.2017.050.

Cayumil, R., Khanna, R., Rajarao, R., Mukherjee, P. S., & Sahajwalla, V. (2016). Concentration of precious metals during their recovery from electronic waste. *Waste Management, 57*, 121–130. https://doi.org/10.1016/j.wasman.2015.12.004.

Chen, P., Miah, M. R., & Aschner, M. (2016). Metals and neurodegeneration. In *F1000Research*, Vol. 5. Faculty of 1000 Ltd. https://doi.org/10.12688/f1000research.7431.1.

Chen, F., Liu, F., Zhou, S., Wang, J., Zeng, Y., & Liao, C. (2023). Selective separation and recovery of selenium and mercury from hazardous acid sludge obtained from the acid-making process of copper smelting plants. *Hydrometallurgy, 221*, 106133. https://doi.org/10.1016/J.HYDROMET.2023.106133.

Chinmay, S. (2022). What materials are used to make solar panels? *AZoM.*

Ciolea, D. I., Ilciuc, O. D., & Berca, M. (2022). Studies and research on the recovery of copper from industrial waste solutions by the cementation method. *Inzynieria Mineralna, 1*(1), 65–70. https://doi.org/10.29227/IM-2022-01-08.

Cleaver, V., Williams, I., & Pierron, X. (2014). The potential for recovering metals from small household appliances. *A Conference Proceeding from the Event Symposium on Urban Mining — Old Monastery of Saint Augustine,* Bergamo, Italy, 19–21 May 2014.

Coman, V., Robotin, B., & Ilea, P. (2013). Nickel recovery/removal from industrial wastes: A review. *Resources, Conservation and Recycling, 73*, 229–238. https://doi.org/10.1016/J.RESCONREC.2013.01.019.

David (2021). How to recover aluminum & precious metals from electronic scrap. *911 Metallurgist*, January 19.

Desmarais, M., Pirade, F., Zhang, J., & Rene, E. R. (2020). Biohydrometallurgical processes for the recovery of precious and base metals from waste electrical and electronic equipments: Current trends and perspectives. *Bioresource Technology Reports, 11*, 100526. https://doi.org/10.1016/J.BITEB.2020. 100526.

Dikshit, P. K., Poddar, M. K., & Chakma, S. (2023). Chapter 12 — Biohydrogen production from waste substrates and its techno-economic analysis. In Antonio Scipioni, Alessandro Manzardo, & Jingzheng Ren (eds.), *Hydrogen Economy,* Second Edition, pp. 399–429. Academic Press. https://doi. org/10.1016/B978-0-323-99514-6.00015-7.

Ding, Y., Zhang, S., Liu, B., Zheng, H., Chang, C. C., & Ekberg, C. (2019). Recovery of precious metals from electronic waste and spent catalysts: A review. *Resources, Conservation and Recycling, 141*, 284–298. https://doi. org/10.1016/J.RESCONREC.2018.10.041.

Dissanayake, V. (2014). Electronic waste. In Philip Wexler (ed.), *Encyclopedia of Toxicology*, Third Edition, pp. 568–572. Academic Press. https://doi. org/10.1016/B978-0-12-386454-3.00565-0.

Evren Tugtaş, A. & Çalli, B. (2018). Removal and recovery of metals by using bio-electrochemical system. In *Microbial Fuel Cell*, pp. 307–333. Springer International Publishing, Cham. https://doi.org/10.1007/978-3-319-66793-5_16.

Free, M. L. (2014). Biohydrometallurgy. In *Treatise on Process Metallurgy,* Vol. 3, pp. 983–993. Elsevier, India. https://doi.org/10.1016/B978-0-08-096988-6.00020-1.

Free, M. L. & Moats, M. (2014). Hydrometallurgical processing. In *Treatise on Process Metallurgy,* Vol. 3, pp. 949–982. Elsevier, India. https://doi. org/10.1016/B978-0-08-096988-6.00021-3.

Gasdia, M. (2023). What happens to the gold after the old jewelry is cashed in? *Thermo Fisher Scientific*, June 5.

Govind, P. & Madhuri, S. (2014). Heavy metals causing toxicity in animals and fishes. *Research Journal of Animal, Veterinary and Fishery Sciences, 2*(2), 17–23.

Gulliani, S., Volpe, M., Messineo, A., & Volpe, R. (2023). Recovery of metals and valuable chemicals from waste electric and electronic materials: A critical review of existing technologies. *RSC Sustainability, 1*(5), 1085–1108. https://doi.org/10.1039/D3SU00034F.

Gupta, D. K., Palma, J. M., & Corpas, F. J. (2015). *Reactive Oxygen Species and Oxidative Damage in Plants under Stress*. Springer, Germany.

Harris, B. (2010). Mineral products and metals that make LED light bulbs. Mec factSheet, U.S. Geological Survey, Society for Mining, Metullargy and Exploration.

Hayati, M., Ganji, S. M. S. A., Shahcheraghi, S. H., & Khabir, R. R. (2023). Optimization of copper recovery from electronic waste using response surface methodology and Monte Carlo simulation under uncertainty. *Journal of Material Cycles and Waste Management, 25*(1), 211–220. https://doi.org/10.1007/s10163-022-01526-2.

Islam, A., Ahmed, T., Awual, M. R., Rahman, A., Sultana, M., Aziz, A. A., Monir, M. U., Teo, S. H., & Hasan, M. (2020). Advances in sustainable approaches to recover metals from e-waste-A review. *Journal of Cleaner Production, 244*, 118815. https://doi.org/10.1016/J.JCLEPRO.2019.118815.

Jadhao, P. R., Ahmad, E., Pant, K. K., & Nigam, K. D. P. (2022). Advancements in the field of electronic waste Recycling: Critical assessment of chemical route for generation of energy and valuable products coupled with metal recovery. *Separation and Purification Technology, 289*, 120773. https://doi.org/10.1016/J.SEPPUR.2022.120773.

Jain, M., Kumar, D., Chaudhary, J., Kumar, S., Sharma, S., & Verma, A. S. (2023). Review on E-waste management and its impact on the environment and society. *Waste Management Bulletin, 1*(3), 34–44. https://doi.org/https://doi.org/10.1016/j.wmb.2023.06.004.

Jerez, C. A. (2009). Metal extraction and biomining. In *Encyclopedia of Microbiology,* Third Edition, pp. 407–420. Academic Press, Elsevier. https://doi.org/10.1016/B978-012373944-5. 00154-1.

Jha, M. K., Choubey, P. K., Jha, A. K., Kumari, A., Lee, J., Kumar, V., & Jeong, J. (2012). Leaching studies for tin recovery from waste e-scrap. *Waste Management, 32*(10), 1919–1925. https://doi.org/10.1016/j.wasman.2012.05.006.

Kahar, I. N. S., Othman, N., Noah, N. F. M., & Suliman, S. S. (2023). Recovery of copper and silver from industrial e-waste leached solutions using sustainable liquid membrane technology: A review. *Environmental Science and Pollution Research, 30*(25), 66445–66472. https://doi.org/10.1007/s11356-023-26951-0.

Karelia, G. (2021). Brothers extract gold & silver from E-waste, save 100000 metric tonnes of carbon. *The Better Mind,* 14 October.

Kiran, R. B. & Sharma, R. (2022). Effect of heavy metals: An overview. *Materials Today: Proceedings, 51*, 880–885. https://doi.org/10.1016/J.MATPR.2021.06.278.

Kostanyan, A. E., Belova, V. V., Zakhodyaeva, Y. A., & Voshkin, A. A. (2023). Extraction of copper from sulfuric acid solutions based on pseudo-liquid membrane technology. *Membranes*, *13*(4), 418. https://doi.org/10.3390/membranes13040418.

Manikandan, S., Inbakandan, D., Valli Nachiyar, C., & Karthick Raja Namasivayam, S. (2023). Towards sustainable metal recovery from e-waste: A mini review. *Sustainable Chemistry for the Environment*, *2*, 100001. https://doi.org/10.1016/j.scenv.2023.100001.

Minay, E. J. & Boccaccini, A. R. (2005). Metals. *Biomaterials, Artificial Organs and Tissue Engineering*, pp. 15–25. Woodhead Publishing, Elsevier. https://doi.org/10.1533/9781845690861.1.15.

Monika, K. J. (2010). E-waste management: As a challenge to public health in India. *Indian Journal of Community Medicine*, *35*(3), 382–385. https://doi.org/10.4103/0970-0218.69251.

Murugappan, R. M. & Karthikeyan, M. (2021). Microbe-assisted management and recovery of heavy metals from electronic wastes. In *Environmental Management of Waste Electrical and Electronic Equipment*, pp. 65–88. https://doi.org/10.1016/B978-0-12-822474-8.00004-0. Elsevier.

Garai, P., Banerjee, P., Mondal, P., & Saha, N. C. (2021). Effect of heavy metals on fishes: Toxicity and bioaccumulation. *Journal of Clinical Toxicology*, *11*(S18), 1–10.

Pant, D., Singh, A., Van Bogaert, G., Irving Olsen, S., Singh Nigam, P., Diels, L., & Vanbroekhoven, K. (2012). Bioelectrochemical systems (BES) for sustainable energy production and product recovery from organic wastes and industrial wastewaters. *RSC Advances*, *2*(4), 1248–1263. https://doi.org/10.1039/C1RA00839K.

Qi, D. (2018). Treatment of wastewater, off-gas, and waste solid. In *Hydrometallurgy of Rare Earths*, pp. 743–777. Elsevier. https://doi.org/10.1016/B978-0-12-813920-2.00008-8.

Reed, B. (2021). What metal is best for agricultural implements? Fairlawn Tool Inc., 9 December.

Robotin, B., Ispas, A., Coman, V., Bund, A., & Ilea, P. (2013). Nickel recovery from electronic waste II Electrodeposition of Ni and Ni–Fe alloys from diluted sulfate solutions. *Waste Management*, *33*(11), 2381–2389. https://doi.org/10.1016/j.wasman.2013.06.001.

Sarangi, C. K., Sheik, A. R., Marandi, B., Ponnam, V., Ghosh, M. K., Sanjay, K., Minakshi, M., & Subbaiah, T. (2023). Recovery of tellurium from waste anode slime containing high copper and high tellurium of copper refineries. *Sustainability*, *15*(15), 11919. https://doi.org/10.3390/su151511919.

Shamsuddin, M. (1986). Metal recovery from scrap and waste. *JOM*, *38*(2), 24–31. https://doi.org/10.1007/BF03257917.

Shulman, A. (2017). How does NASA use aluminum? *Avion Alloys*, 24 March.

Singh, M. Y. (2021). Is it E-waste or gold, silver, and platinum waste? *Electronics B2b.Com*, 6 November.

Singh, J. & Kalamdhad, A. (2011). Effects of heavy metals on soil, plants, human health and aquatic life. *International Journal of Research in Chemistry and Environment*, *1*, 15–21.

Singh, S. P. & Schwan, A. L. (2011). Sulfur metabolism in plants and related biotechnologies. In *Comprehensive Biotechnology*, Second Edition, Vol. 4, pp. 257–271. https://doi.org/10.1016/B978-0-08-088504-9.00268-3.

Singh, R., Gautam, N., Mishra, A., & Gupta, R. (2011). Heavy metals and living systems: An overview. *Indian Journal of Pharmacology*, *43*(3), 246–253. https://doi.org/10.4103/0253-7613.81505.

Sivakumar, P., Prabhakaran, D., & Thirumarimurugan, M. (2022). Optimization studies on E-waste for the recovery of zinc and aluminium by electro deposition. *Journal of Civil and Environmental Engineering*, *12*, 6. https://doi.org/10.37421/2165-784X.

Sree Lakshmi, V., Rahul Satya, C. H., & Aravind, D. N. (2021). Electronic waste: Overview, recycling and metal extraction methods. *IOP Conference Series: Materials Science and Engineering*, *1136*(1), 012069. https://doi.org/10.1088/1757-899X/1136/1/012069.

Tchounwou, P. B., Yedjou, C. G., Patlolla, A. K., & Sutton, D. J. (2012). Heavy metal toxicity and the environment. *Experientia Supplementum*, *101*, 133–164. https://doi.org/10.1007/978-3-7643-8340-4_6.

Tripathee, L., Kang, S., Li, C., Sun, S., & Sharma, C. M. (2019). Chemical components and distributions in precipitation in the Third Pole. In *Water Quality in the Third Pole: The Roles of Climate Change and Human Activities*, pp. 3–41. https://doi.org/10.1016/B978-0-12-816489-1.00001-3.

Ulbrich Stainless Steels & Special Metals, Inc. (2021). Finding the right metals and alloys for medical device manufacturing. *AZoM*, 4 March.

Valix, M. (2017). Bioleaching of electronic waste: Milestones and challenges. In *Current Developments in Biotechnology and Bioengineering: Solid Waste Management*, pp. 407–442. https://doi.org/10.1016/B978-0-444-63664-5.00018-6.

Vending Machine (n.d.). Made how.

Vera, M., Schippers, A., Hedrich, S., & Sand, W. (2022). Progress in bioleaching: Fundamentals and mechanisms of microbial metal sulfide oxidation - part A. *Applied Microbiology and Biotechnology*, *106*(21), 6933–6952. https://doi.org/10.1007/s00253-022-12168-7.

Xia, J., Mahandra, H., & Ghahreman, A. (2021). Efficient gold recovery from cyanide solution using magnetic activated carbon. *ACS Applied Materials & Interfaces*, *13*(40), 47642–47649. https://doi.org/10.1021/acsami.1c13920.

Xing, P., Wang, C., Wang, L., Ma, B., & Chen, Y. (2019). Hydrometallurgical recovery of lead from spent lead-acid battery paste via leaching and electrowinning in chloride solution. *Hydrometallurgy, 189*, 105134. https://doi.org/10.1016/J.HYDROMET.2019.105134.

Yahya, F. N., Ibrahim, W. H. W., Aziz, B. A., & Suli, L. N. M. (2020). Simulation of leaching process of gold by cyanidation. *IOP Conference Series: Materials Science and Engineering, 736*(2), 022111. https://doi.org/10.1088/1757-899X/736/2/022111.

Yu, Z., Han, H., Feng, P., Zhao, S., Zhou, T., Kakade, A., Kulshrestha, S., Majeed, S., & Li, X. (2020). Recent advances in the recovery of metals from waste through biological processes. *Bioresource Technology, 297*, 122416. https://doi.org/10.1016/J.BIORTECH.2019.122416.

Zhang, S., Forssberg, E., Arvidson, B., & Moss, W. (1998). Aluminum recovery from electronic scrap by High-Force® eddy-current separators. *Resources, Conservation and Recycling, 23*(4), 225–241. https://doi.org/10.1016/S0921-3449(98)00022-6.

Chapter 3

Recent Advancements in the Valorization of Used Lithium Ion Battery

**Uttam Bista*, Swapan Suman†, Rajendra Joshi‡,
and Dilip Kumar Rajak§**

*Department of Applied Sciences and Chemical Engineering,
Pulchowk Campus, Institute of Engineering (IoE),
Tribhuvan University, Kathmandu, Nepal*

†*Mechanical Engineering Department, Meerut Institute of
Engineering and Technology, Meerut, India*

‡*Department of Pharmacy, Kathmandu University, Dhulikhel, Nepal*

§*Department of Chemical Science and Engineering,
Kathmandu University, Dhulikhel, Nepal*

§*dilip.123bit@gmail.com*

Abstract

Used lithium-ion batteries (LIBs), which contain valuable metals such as cobalt, lithium, manganese, and nickel, are a promising source for future lithium production, with recycling efforts underway. However, lithium recovery is difficult, making LIBs a better source compared to seawater. LIBs contain graphite-coated copper and $LiCoO_2$-coated aluminum electrodes bound with polyvinylidene fluoride (PVDF), serving as conductors through lithium salt-based electrolytes, requiring physical separation and metal enrichment before hydrometallurgical treatment. Flammable electrolytes require discharge and immersion in specific solutions. Hydrometallurgical processes can increase metal concentrations but emit hazardous gases, making them less environmentally friendly. Mechanochemical methods, such as dry and wet milling, can improve lithium and cobalt recovery by inducing structural changes in electrode materials, increasing extraction efficiency, and improving the separation of magnetic materials. Pretreatment techniques for extracting lithium and cobalt emphasize hydrometallurgical recovery through leaching and impurity removal, with a particular focus on the unique challenges posed by lithium extraction due to its stability in aqueous solutions. The economic factors of recycling LIBs depend on recycling rates and metal prices, with hydrometallurgical processes offering the lowest energy consumption and the potential for lithium recovery.

Keywords: Valorization, lithium, used libs, hydrometallurgical recovery.

1. Introduction

The global expansion of clean and renewable energy is crucial for sustainability, and lithium-ion batteries (LIBs), with their unique technical characteristics, have the potential to transform the renewable energy market, particularly in applications such as plug-in hybrid automobiles, electronics goods, the military, and healthcare sectors, driving down LIBs costs for off-grid renewable energy storage (Choubey *et al.*, 2021; Zhou *et al.*, 2023).

Lithium's economic and strategic importance, along with its classification as an "energy-critical element," underscores its vital role in LIBs. Increased metallurgical recovery of lithium from various sources is necessary due to anticipated shortages in the lithium market, which are a result of factors such as Tesla Gigafactory's plans for massive electric vehicle production and the introduction of cheaper LIBs (Hua *et al.*, 2021; Panda *et al.*, 2023). Although conventional resources such as lithium production from minerals and brine are available, they require significant investments, and the global availability of primary lithium resources is limited. However, the potential use of seawater, which boasts vast reserves compared to minerals and brine, can alleviate this issue. Furthermore, the proliferation of LIBs in various rechargeable devices generates a significant amount of waste containing lithium, serving as a substantial post-usage lithium reservoir (Wang *et al.*, 2022).

These non-conventional lithium resources are not geographically restricted since sea water is widely available, allowing countries using LIBs to capitalize on their spent batteries as a potential lithium resource through proper recycling methods. However, processing sea water presents challenges due to its low lithium content, and dealing with impurities in aqueous solutions from treated LIBs further complicates downstream processing (Kamran *et al.*, 2019). Therefore, reviewing and improving existing processing methods is crucial for enhancing extraction and lithium recovery both economically and technologically. This book chapter discusses diverse upstream beneficiation processes for spent LIBs aimed at augmenting metal concentrations within LIBs. Additionally, it explores upstream extraction methodologies, with a focus on environmental and socioeconomic considerations.

2. Methods for Recovering Lithium from Depleted LIBs

Spent LIBs exhibit a notable richness in critical elements, specifically cobalt (5–33 wt. %), manganese (15–20 wt. %), lithium (5–7 wt. %), and nickel (0.02–0.3 wt. %). They hold promise as a future source for lithium retrieval (Garole *et al.*, 2020; S. Wang *et al.*, 2020). Cobalt, a relatively rarer element than lithium, is recuperated at a higher rate relative to lithium. The elevated lithium concentration (5–7 wt. %) in discarded LIBs renders them more valuable when juxtaposed with seawater (0.1–0.2 ppm) (Bowell *et al.*, 2020; Kamran *et al.*, 2019). Significant endeavors have been undertaken in this domain to recycle LIBs and extract their constituent metals. Owing to the paucity of impurities in spent LIBs, a limited number of commercial techniques are sufficient. This stands in contrast to the treatment of their mineral counterparts, which necessitate substantial pre-processing energy inputs, as well as the treatment of brines and seawater. Consequently, depleted LIBs emerge as the primary source of lithium in this investigation, with a delineation of the lithium recovery process provided. Figure 1 shows the concise procedural stages in the extraction of lithium from spent LIBs.

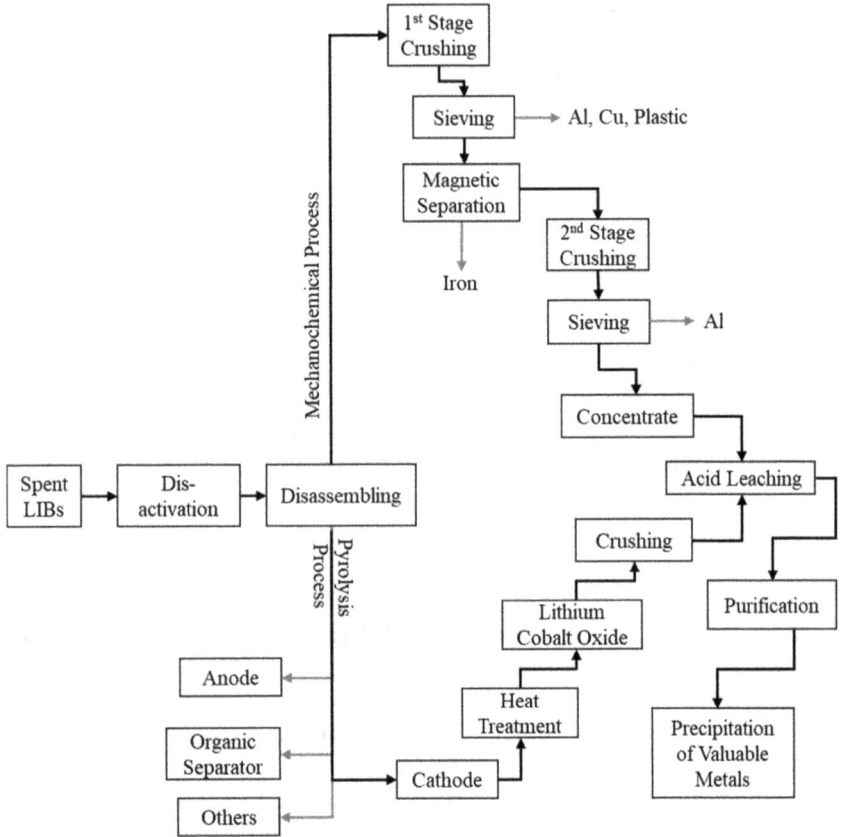

Figure 1. Basic flowsheet for metal recovery process from spent LIBs. (Adapted and modified from Choubey *et al.*, 2017).

2.1 *Dismantling and deactivating spent LIBs*

LIBs comprise negative ($LiCoO_2$ on aluminum foil) and positive (graphite on copper) electrodes, a separator, an electrolyte, and a stainless-steel housing. Butadiene-styrene polyvinylidene fluoride (PVDF), copolymers, or modified cellulose support the graphite and mixed oxide particles. Lithium-salt liquid electrolytes in organic solvents facilitate current conduction between electrodes in an external circuit. Prior to hydrometallurgical processing, LIBs require physical disassembly, followed by enrichment of the resultant black mass containing relevant metals.

Flammable electrolytes persist in spent LIBs, necessitating pre-discharge to avert short circuits (Fu *et al.*, 2020). A common deactivation method involves immersion in salt solutions, distilled water (possibly with Fe powder), or liquid nitrogen (Mossali *et al.*, 2020; Wang *et al.*, 2018). Caution is advised when applying pressure to closed-condition LIBs, as it may induce an internal short circuit.

2.2 Black mass pretreatment and enrichment

Following discharge, a sequence of processes — comprising crushing, sieving, magnetic separation, fine crushing, classification, and employment of a magnetic separator — effectively eliminate steel casing remnants while concentrating cobalt and lithium within the black mass.

2.2.1 Mechano-chemical and mechanical processing

Hydrometallurgical processes can increase the metal concentration in mechanically treated used LIBs. This makes cleaning the leachate less important. This entails multistage pulverization and sifting for metal value enrichment. However, during mechanical processing, the cathode's robustness emits hazardous gases (Garole *et al.*, 2020). Achieving precise component separation is challenging due to the complexities associated with the presence of combined metals and inorganic and organic subtance. Garole *et al.* (2020) contend that mechanical treatment lacks environmental friendliness for spent LIB recycling. Pinegar and Smith (2020) state that hot N-methyl pyrrolidone (NMP) and mechanical crushing and grinding are effective methods of removing binders and supporting substrates. Regarding mechanochemical lithium recovery, studies have explored both dry and wet milling methods (Alavi *et al.*, 2022).

$LiCoO_2$ can be grounded with polyvinyl chloride (which serves as a chloride source) in a planetary ball mill to produce lithium and cobalt chlorides that are readily extracted via water leaching in the presence of air (Cheng *et al.*, 2023). It was found that the crystal structure of $LiCo0.2Ni0.8O_2$ can be altered through mechano-chemical treatment with or without alumina. Mixing alumina transformed the crystalline structure into an amorphous one, thereby increasing metal extraction efficiency to over 90% (Wang *et al.*, 2020). It was discovered that dry milling of spent LIBs is more effective for separating magnetic materials from the steel casing. This aided in the concentration of the targeted metals within the black mass, thus increasing the extraction efficiency during the dissolution phase.

2.2.2 Pyrolysis and heat treatment

LIBs contain several organic binders which may inhibit liquid–liquid metal separation from leach fluid. Lithium sorption on graphite surfaces reduces lithium recovery (Wang *et al.*, 2020). Therefore, graphite and organic compounds must be thermally pretreated before leaching.

A thermal treatment study was conducted by Murali and coworkers, who used a muffle furnace to heat LIB samples to 100–150°C before disassembling them using a high-speed shredder. Second, the shreds were heated in a furnace and vibrated to separate electrode materials. $LiCoO_2$ was derived by removing carbon and binder at temperatures between 500 and 900°C for 30–120 minutes (Murali *et al.*, 2021). Pretreatment also involved depositing nitric acid slurry for $LiCoO_2$ leaching in a stainless-steel crucible and calcining it at 500–1000°C in air. In a muffle furnace, the active mass of LIBs was mixed with $KHSO_4$ which weighed eight times more than LIBs for 3 hrs at 500°C (Zhuang *et al.*, 2019). To minimize sulfate conversion into SO_2 and sulfide species, the heating process was conducted with airflow. Calcined active material at 500°C for 300 minutes improves lithium extraction through water dissolution. This was because carbon is a sorbent, and calcination eliminates it. Garole *et al.* (2020) observed that burning used LIB active matter at 900°C reduced cobalt leaching recovery without affecting lithium dissolving in H_2SO_4 solution. Takacova *et al.* (2023) found that cobalt dissolution and lithium extraction in H_2SO_4 improved when the temperature of incineration was reduced to 700°C. He *et al.* (2021) performed the thermal decomposition of polyvinylidene fluoride (PVDF).

Adhesion neutralization via incineration and impact liberation (ANVIIL) is the process of removing adhesion from PVDF binder materials by combining thermal treatment and mechanical separation. Thermogravimetric analysis at 350–650°C determined the best PVDF decomposition temperature. The polymeric PVDF binder decomposed at 350°C in oxygen. In 20 minutes, nearly 98.85% of the binder mass evaporated at 550°C. Temperature increases up to 650°C caused carbon black loss and graphite loss. Optimal PVDF decomposition occurs at approximately 500–580°C, inhibiting the deterioration of other compounds.

The current battery binder calcination process emits dangerous combustion fumes, which is its main problem. Yang *et al.* (2020) reported that the thermal treatment procedure completely decomposes PVDF at 550°C, which is below the melting point of aluminum (Al) foils (650°C). The cathode conductor acetylene black and the active cathode materials $LiNi_{1/3}Co_{1/3}Mn_{1/3}O_2$ undergo redox reactions that change their charge from high to low through heat treatment. During leaching, this enables the recovery of a large amount of metal.

Heat treatment of used LIBs is easy and practical. Processing costs rise due to the need for equipment to remove carbon and organic compounds from off-gas emissions and smoke. In light of these drawbacks, researchers have recently favored pyrolysis over mechanical and thermal treatments of spent LIBs. Zhou *et al.* (2023) separated cathode active material from Al foils using pyrolysis at 600°C, since most binders evaporated or dissolved in this process. Al foils become brittle beyond 316°C, making cathode active material separation difficult. This effect worsens above 700°C. Due to adhesive compounds and cathode production methods, cathode-active materials can preserve the integrity of Al foils; however, flaking is unavoidable. The advantages of vacuum pyrolysis over mechanical treatment are evident (Yazdani *et al.*, 2019). Metal oxidation can be limited to reduce harmful fumes and organic compound breakdown. Thus, cobalt, lithium, and the pyrolysis products from the electrolyte and binder can be efficiently recycled. FT-IR examination of pyrolyzed products shows that most condenser PVDF materials can be recycled (Ahamed *et al.*, 2023). After pretreatment, the hydrometallurgical process can be used to recover lithium in addition to the more expensive cobalt.

2.2.3 *Ultrasonic flossing*

As an alternative to physical and mechanical pre-treatment methods, some researchers have used ultrasonic treatment in solvents such as NMP, N-N-dimethylformamide (DMF), N-N-dimethylacetamide (DMAC); N-N-dimethyl sulfoxide (DMSO), and ethanol to separate the cathode material. Under these conditions, the binder materials (PVDF) are dissolved at 60°C using 240 W of ultrasonic power for 30 minutes, with a liquid–solid ratio of 10/1. According to the "like dissolves like" principle, a polar compound is readily dissolved in a polar solvent. Figure 2 shows that NMP, a strong polar solvent, dissolves PVDF well and has the highest cathode peel-off efficiency, while ethanol, a weak polar solvent, barely

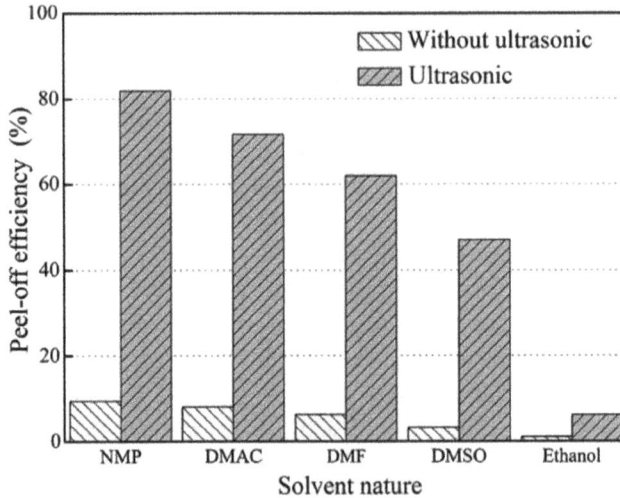

Figure 2.   Cathode peel-off efficiency by solvent. (Adapted from He *et al.*, 2015).

dissolves it (Zhang *et al.*, 2019). Two distinct mechanisms can explain the detachment of cathode materials from Al. First, ultrasound speeds up the convective movement of the solvent. This makes it easier for PVDF to dissolve and peel off the cathode material (Zhang *et al.*, 2019). Ultrasonic cavitation is the second phenomenon (Ding *et al.*, 2023; Torkashvand *et al.*, 2021; Waghmare *et al.*, 2019). Acoustic pressure cycles between compression and rarefaction when the ultrasonic power is supplied. Ultrasonic waves may create vacuum cavities in the solvent during rarefaction because the solvent molecules' pressure drops.

As gases dissolved in the solvent enter these vacuum holes, millions of cavitation bubbles form and expand. Figure 3 shows how cavitation bubbles burst during compression, causing a strong impact force at the cathode–solvent contact. This separates the cathode material from the Al foils. Thus, PVDF dissolution and ultrasound cavitation together cause the cathode material to peel off from the Al foils.

2.2.4 *Chemical pre-treatment*

Before extracting lithium, a pre-treatment technique that includes alkali leaching is regarded as necessary in the recycling of spent LIBs in order to remove impurities such as Al and copper (Cu). Cheng *et al.* (2023) and Takahashi *et al.* (2020) reported the use of caustic leaching to remove Al

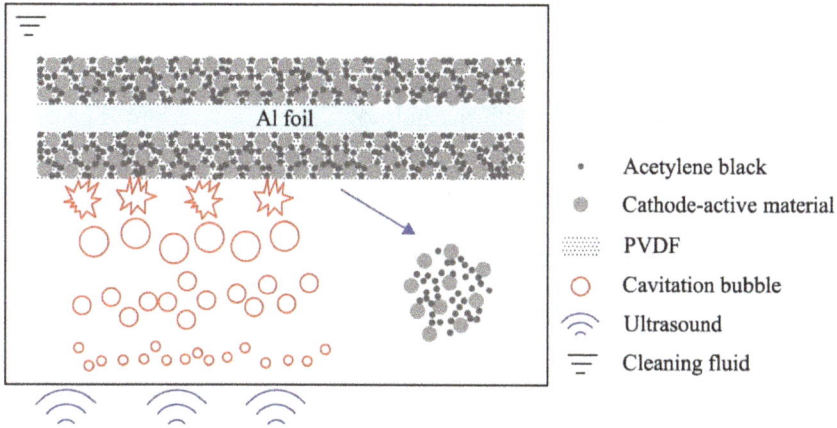

Figure 3.   Ultrasonic cleaning for cathode material and aluminum recovery from spent LiBs. (Adapted from He *et al.*, 2015).

from cathode materials in a specific way. Equations (1) and (2) represent the reactions occurring during the dissolution of Al with NaOH:

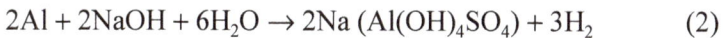

$$Al_2O_3 + 2NaOH + 3H_2O \rightarrow 2Na\,(Al(OH)_4SO_4) \tag{1}$$

$$2Al + 2NaOH + 6H_2O \rightarrow 2Na\,(Al(OH)_4SO_4) + 3H_2 \tag{2}$$

Caustic solutions do not form soluble compounds with cobalt and lithium; therefore, they remain as residues during reactions. At 50°C and a 100 g/L pulp density (PD), 15 wt. % NaOH leaches 60% Al and some lithium from the pulp. A 240 minute leaching time with 5 wt. % NaOH could increase Al leaching to 99.9%, but it is not advised. After selectively removing Cu and Al impurities, acids can leach out the residues. Before leaching lithium and cobalt, Nayl *et al.* (2017) utilized ammonia solution (NH_3OH) to remove Cu and Al. With 4 mol/L NH4OH at 80°C, a 66.6 g/L PD, and a 60 minute reaction time, 98% Al, 65% Cu, and very little Li, Co, and Mn were extracted.

3. Aqueous Extraction of Valuable Metals from Spent LIBs

Many approaches were applied to extract valuable metals from aqueous solutions of different media following the previous methods, which aimed at extracting more metals from used LIBs. Cobalt was traditionally the main metal extracted (Butt *et al.*, 2022; Wang *et al.*, 2020). After the pre-treatment of used LIBs, leaching is the most important step for hydro-metallurgical recovery of the desired metal(s), achieved by immersing them in aqueous solutions with the help of a suitable lixiviant. Leaching experiments have used mineral acids, organic acids, and alkaline solutions.

3.1 *Mineral acids leaching*

Mineral acids are frequently used as lixiviants to cleanse spent LIBs for cobalt and lithium extraction due to their aggressive nature. Table 1 presents a selection of mineral acid leaching outcomes observed on spent materials. Wang *et al.* (2021) and Zheng *et al.* (2018) suggest acid leaching with crushing and ultrasonic washing to remove $LiCoO_2$ from used LIBs instead of physical treatment. A 50–120 minute treatment at 90°C is sufficient for the HCl leaching of spent LiBs to recover more than 95% Li and Co (Ilyas *et al.*, 2022). Equation (3) shows the reaction:

$$2LiCoO_2 + 6HCl \rightarrow 2CoCl_2 + 2LiCl + 3H_2O + 0.5O_2 \qquad (3)$$

With 2 mol/L HCl and H_2O_2 as a reducing agent, the lithium extraction rate dropped to 80% after 3 h. The same reagent needs 18 h at a higher temperature to increase leaching efficiency (producing more than 80% lithium). Additionally, the efficiency may be improved by leaching the unleached residue again. To produce a valuable lithium carbonate salt, the leach fluid is purified numerous times to remove elements such as Co, Ni, and Li. At 80°C and 100 g/L, Yang and co-workers achieved leaching of over 99% Li in 60 minutes with a solution of 1 mol/L HCl and 4 vol% H_2O_2. For the same leaching action with H_2SO_4 and H_2O_2, leaching for 90 minutes at 90°C is required (Sattar *et al.*, 2019). HCl leaching is likely to be fast because chloride ions disrupt the surface layer. According to reaction (4), *in situ* generation of chlorine gas may also improve extraction:

$$2LiCoO_2 + 8HCl \rightarrow 2CoCl_2 + Cl_2 + 2LiCl + 4H_2O \qquad (4)$$

$$H_2O_2 \rightarrow H_2O + 0.5O_2 \qquad (5)$$

Cl_2 is a highly hazardous gas, requiring the use of specialized equipment, which raises the price of recycling and poses serious environmental

Table 1. Typical lixiviants for spent LiBs.

Lixiviant	Operating condition	Metal recovery	Pros	Cons	References
1.75 M HCl	S/L: 200 g/L, 50°C for 50 min	More than 90 % recovery of Co, Mn and Li	Effective leaching in a brief duration	Lack of exactness	Ilyas et al., 2022
1 M HCl/ H_2O_2 (4.0 vol. %)	S/L: 100 g/L, 80°C for 60 min	Li: 100%, Cu: 98.5%, and Al: 99.2%	HCl is a highly aggressive leachate	Effort is needed for the selective recovery of Li	Yang et al., 2019
2 M H_2SO_4/ H_2O_2 (4.0 vol. %)	S/L: 50 g/L, 90°C for 90 min	Li: 98%, Ni: 99%, Co and Mn: 94%	H_2O_2 addition markedly sped up metal dissolution	Leaching time is relatively high	Sattar et al., 2019
0.7 M H_3PO_4/ H_2O_2 (4.0 vol %)	S/L: 50 g/L, 40°C for 60 min	Li and Co: 99 %	Shorter reaction duration	Recovery of unused H_3PO_4 is required	Chen et al., 2017
1.53 M acetic acid (CH_3COOH)/ glucose ($C_6H_{12}O_6$) (167 g/g)	S/L: 5.19 g/L, 185°C for 67 min	More than 97% recovery of Li, Co, Mn and Ni	Reuse of reagents and secondary extraction of residues	High reaction temperature	Liang et al., 2022
1.5 M citric acid ($C_6H_8O_7$)/ H_2O_2 (6.0 vol %)	S/L: 15 g/L, 90°C for 90 min	Li: 97% and Co: 99.5%	Citric acid is nature-friendly and eco-conscious	Leaching temperature is high	Xu et al., 2021
0.2 M L-tartaric acid ($C_4H_6O_6$)/ H_2O_2 (6.0 vol %)	S/L: 15 g/L, 90°C for 90 min	More than 95% recovery of Li and Co	Less leaching time and eco friendly	Development of scale-up is still under process	
1.5 M succinic acid ($C_4H_6O_4$)/ H_2O_2 (4.0 vol %)	S/L: 15, 70°C for 40 min	Li: 96% and Co: 99.5%	High leaching efficiency of Li and Co	Leachate concentration is high	Li et al., 2015

dangers when corrosion-resistant machinery is not in place. Takahashi *et al.* (2020) cite the sulfate leaching system as the preferred environmentally friendly method. Using sulfuric acid dissolution to recycle spent LIBs, K. Wang *et al.* (2021) used 2 vol. % H_2O_2 as a cobalt reductant to speed up the reaction. $LiCoO_2$ from exhausted LIBs was leached in 2 mol/L H_2SO_4 at 33 g/L PD and 60°C for 120 minutes to remove 87.5% Li and 96.3% Co. As shown in Equation (5), temperatures above 60°C may decompose H_2O_2 into water; hence, 60°C was chosen for maximal leaching efficiency.

The reductant assists in dissolving cobalt by converting the less soluble Co^{3+} species to the more soluble Co^{2+} species. Since reductants have no value. Lithium produces only Li^+ in water. In recent studies, spent LIBs were enriched via vacuum pyrolysis rather than mechanical treatment (Choubey *et al.*, 2021; Maroufi *et al.*, 2020; Zhou *et al.*, 2023). After 60 minutes at 5 vol. % H_2O_2 and 80°C, 2 mol/L H_2SO_4 leached 50 g/L of pyrolyzed material with 99% lithium and cobalt extraction efficiency. Lithium leaching remained unaffected up to 100 g/L PD, while extraction of cobalt decreased because stoichiometric acid was less available than lithium. $LiCoO_2$ leaches with mineral acids (H_2SO_4, HNO_3, and HCl) in H_2O_2 (Choubey *et al.*, 2021; Maroufi *et al.*, 2020; Zhou *et al.*, 2023):

$$LiCoO_2 + 0.5H_2O_2 + 0.5H_2SO_4 \rightarrow CoSO_4 + 0.5Li_2SO_4 + 3H_2O + O_2 \quad (6)$$

$$2LiCoO_2 + H_2O_2 + 6HCl \rightarrow 2CoCl_2 + 2LiCl + 4H_2O + O_2 \quad (7)$$

$$2LiCoO_2 + H_2O_2 + 6HNO_3 \rightarrow 2Co(NO_3)_2 + 2LiNO_3 + 4H_2O + O_2 \quad (8)$$

Recently, Meshram *et al.* (2023), Sun *et al.* (2021), and Zhao *et al.* (2020) examined sodium bisulfite as a viable H_2O_2 alternative for LIB leaching. The authors have shown that H_2O_2 dissolves several metals in a similar way. Interestingly, lithium extraction was similar with or without the reductant. Leaching efficiency without sodium bisulfite was 93.4% for Li, 66.2% for Co, 96.3% for Ni, and 50.2% for Mn at 1 mol/L H_2SO_4, 95°C, and 240 minutes. At a PD of 20 g/L and a reducing agent of 0.075 mol/L sodium bisulfite, the efficiency of leaching for Li was 96.7%, Co was 91.6%, Ni was 96.4%, and Mn was 87.9%. Around 99% Li and Co recovery was also achieved using 0.7 M H_3PO_4 and 4.0 vol % H_2O_2 as a lixiviant for waste LiBs (Chen *et al.*, 2017).

3.2 *Organic acids leaching*

Mineral acid leaching is a well-established process for recovering metals from spent LIBs; however, it releases hazardous gases such as NO_x, Cl_2, and SO_3. Thus, several researchers are studying metal recovery by organic acid leaching (Table 1). Acetic acid (CH_3COOH), malic acid ($C_4H_6O_5$), oxalic acid ($H_2C_2O_4$), succinic acid ($C_4H_6O_4$), aspartic acid ($C_4H_7NO_4$), L-tartaric acid ($C_4H_6O_6$), and citric acid ($C_6H_8O_7$) have been extensively used for this purpose (Li *et al.*, 2015; de Oliveira Demarco *et al.*, 2019; Ding *et al.*, 2023; JenniLie & ChernLiu, 2021; Xu *et al.*, 2021; K. Wang *et al.*, 2021, 2022; Liang *et al.*, 2022). Either 1.5 mol/L $C_4H_6O_5$ with 2 vol. % H_2O_2 or 1.25 mol/L $C_6H_8O_7$ with 1 vol. % H_2O_2 could leach off all lithium and 90% of cobalt at 90°C and 20 g/L PD. Due to its highest conductivity at 100°C and ion-transfer rate, increasing the malic acid concentration above 2 mol/L does not affect leaching performance. The malic acid leaching reaction is

$$2LiCoO_2 + H_2O_2 + 6C_4H_6O_5 \rightarrow 4LiC_4H_5O_5 + 2Co(C_2H_5O_5) + 4H_2O + O_2 \quad (9)$$

Ding *et al.* (2023) found that citric acid was more effective than other mineral acids (HCl and H_2SO_4) when ultrasonic leaching was used for lithium extraction from discarded LIBs. Under the conditions of 2 mol/L $C_6H_8O_7$, 0.55 mol/L H_2O_2, 25 g/L PD, 60°C, 300 min, and, most crucially, 90 W ultrasonic power, practically all lithium was leached with 96% co-extracted cobalt.

Ascorbic acid leaches depleted LIBs better than citric and malic acids. Even with a higher PD (25 g/L), leaching 98% Li and Co with 1.25 mol/L ascorbic acid takes 20 minutes at 70°C. Since ascorbic acid is a self-reductant, it does not require additional reducing agents, specifically for leaching cobalt. This technology could be investigated on a larger scale and used commercially. The leaching reaction (Jenni Lie & Chern Liu, 2021) can be expressed as follows:

$$2LiCoO_2 + 4C_6H_8O_6 \rightarrow C_6H_6O_6 + C_6H_6O_6Li_2 + 2C_6H_6O_6Co + 4H_2O \quad (10)$$

Cheng *et al.* (2023) and Yan *et al.* (2021) used a strong organic acid (e.g., oxalic acid) to break the strong chemical bond between Li and Co in $LiCoO_2$. By adding H_2O_2, the rate of dissolution could be increased due to a higher rate of replacement of Li^+ ions of $LiCoO_2$ with H^+ ions of oxalic acid, while Co^{3+} is reduced to Co^{2+}. Under the following conditions, nearly 98% of lithium and cobalt were extracted: 1.0 mol/L oxalic acid and 50 g/L PD at 80°C for 2 h. The benefits of this organic acid leaching system include the selective precipitation of cobalt as oxalate salt (CoC_2O_4), while Li and Al remain in the leach liquor and can be subsequently processed using precipitation methods for Li recovery. The reactions of oxalic acid leaching with lithium cobalt oxide are represented by the following equations:

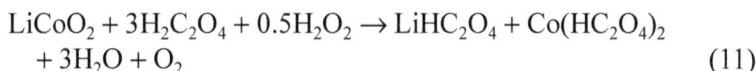

$$LiCoO_2 + 3H_2C_2O_4 + 0.5H_2O_2 \rightarrow LiHC_2O_4 + Co(HC_2O_4)_2 + 3H_2O + O_2 \tag{11}$$

$$2LiCoO_2 + 3H_2C_2O_4 + H_2O_2 \rightarrow Li_2C_2O_4 + 2CoC_2O_4 + 4H_2O + O_2 \tag{12}$$

Wang *et al.* (2022) used the straightforward biodegradation of succinic acid leaching from used LIBs to extract lithium. The efficiency of lithium leaching is 96%, together with all cobalt in the leach fluid, under the conditions of 1.5 mol/L succinic acid, 4 vol. % H_2O_2, and 15 g/L PD at 70°C for up to 40 minutes, which is encouraging. The succinic acid leaching is shown in Equation (13):

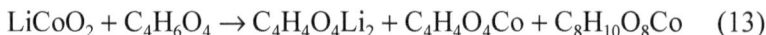

$$LiCoO_2 + C_4H_6O_4 \rightarrow C_4H_4O_4Li_2 + C_4H_4O_4Co + C_8H_{10}O_8Co \tag{13}$$

When aspartic acid ($C_4H_7O_4N$) was used as a lixiviant, lithium and cobalt were recovered at lower rates than those achieved with the aforementioned organic acids. In addition to its diminished water solubility, aspartic acid is a weak acid, making it an ineffective lixiviant for spent LIBs (de Oliveira Demarco *et al.*, 2019). It was found that just 60% of Li and Co can be leached in 1.5 mol/L, L-aspartic acid with 4 vol. % H_2O_2 with 10 g/L PD at 90°C for 2 h (Wang *et al.*, 2020).

4. Separating and Purifying Leach Liquors

Leach liquors from LIBs are separated and processed to eliminate metal ions. Two sections make up the impurity removal steps operationally. Mn, Cu, Al, and Fe are generally removed before isolating Ni and Co from Li-containing solutions. Often, this is done considering the stability order, the hydrolysis constant, coordination chemistry, the function of the d-orbital in metal complexation, the thermodynamics of a reaction, and so on. The utilization of solvent extraction for lithium extraction from LIB leach solutions is not recommended due to these characteristics. The stability of hydrated lithium ions surpasses that of water, resulting in the retention of lithium in aqueous solutions. Thus, Li extraction from aqueous solutions involves more basic extractants than water (Murali *et al.*, 2021). As discussed earlier, solvent synergy is usually obtained at a higher equilibrium pH (>6). Lithium remains in the raffinate, but Al, Co, Ni, Cu, Mn, and Fe can be easily extracted from pH 6.5 LIB leach solutions.

4.1 *Al, Cu, Mn, and Fe separation*

4.1.1 *Through solvent extraction*

To investigate the separation process, a leach liquor containing Co (16.9 g/L), Li (3.8 g/L), Fe (0.6 g/L), Ni (0.15 g/L), Al (0.7 g/L), and Cu (0.4 g/L) was used along with the organic solvents Ionquest 801, Acorga M5640, and their synergistic composition (Pranolo *et al.*, 2010). At equilibrium pHs of 2.5 and 4.0, respectively, Acorga M5640 (2 vol. %) was utilized to extract 99.9% Cu and 55.5% Fe. Using 7 vol. % Ionquest 801 at pHs 2 and 5, respectively, we extracted total Fe and Al. A higher pH of 6 allows the extraction of 60% Cu and 10% Co. It means that neither Acorga nor Ionquest can entirely separate Cu, Al, and Fe.

To extract Cu, Al, and Fe without Co, Acorga M5640 (2%) and Ionquest 801 (7%) were mixed. Due to Acorga's copper affinity, the organic mixture lowered the copper extraction isotherm (pH50) from 5.45 to 2.0, which is substantially lower than that of Ionquest 801. Cu, Al, and Fe separated completely from Co, Ni, and Li at pH 4.5. After Cu removal using Acorga M5640 at pH 1.5–2.0, Al extraction using PC-88A at pH 2.5–3.0 is performed (Ahamed *et al.*, 2023).

4.1.2 *Through chemical precipitation*

Removing metal impurities such as Fe, Cu, Mn, and Al is one of the easiest processes, depending on the solution's pH and hydrolysis constant. Diagrams of the potential and solubility of metal hydroxide species as functions of pH are useful for this (Pourbaix, 1966), as they reveal plausible mechanisms for the precipitation-based separation of Fe, Cu, Al, and Mn at a particular pH. Using a diluted NaOH and $CaCO_3$ solution. Wen & Lee (2023) adjusted the pH to 6.5 and precipitated the hydroxides of Cu, Al, and Fe. In contrast, Cheng *et al.* (2023) precipitated each of the metal impurities in stages. The optimal pH values for the removal of Fe, Mn, and Cu were 3.1 at 95°C, 4.0 at 70°C, and 5.5 at 30°C, respectively. In this investigation, $(NH_4)2S_2O_8$ served as the precipitating agent for Mn (Cheng *et al.*, 2023), which was modified by precipitating MnO_2 at pH 2.0 with $KMnO_4$ as the precipitating reagent (Yang *et al.*, 2020). Choubey *et al.* (2021) optimized the dosage in molar concentration as MnO_2:$KMnO_4$ = 1:2, while Ning *et al.* (2020) suggested heating the solution to 50°C to achieve the desired level of Mn precipitate. Many studies have shown that precipitation removes contaminants and loses Co and Li. MnO_2 is employed as a lithium adsorbent, and its increased surface area and *in situ* precipitated iron hydroxides allow cobalt precipitation and sorption. (Ochromowicz *et al.*, 2021; Srivastava *et al.*, 2013). Thus, lithium and cobalt losses result from sorption onto Fe and Mn precipitates. At pH 7.5, a saturated Na_2CO_3 solution precipitated $MnCO_3$, leaving Ni, Co, and Li in solution (Nayl *et al.*, 2017).

4.2 Ni and Co separation

After separating Fe, Cu, Mn, and Al from the leach solution of LIBs, Co and Ni must be removed before lithium values can be recovered. The process of removing Co and Ni can be further divided into two processes: solvent extraction and chemical precipitation, which are explained in the following sections.

4.2.1 Solvent extraction

Ahamed *et al.* (2023) performed a series of solvent extraction experiments using a 0.35 M sulfuric acid aqueous solution containing copper sulfate, lithium sulfate, cobalt sulfate, manganese sulfate, nickel sulfate, and aluminum sulfate (III). Various extractants, namely Cyanex-272, Acorga M5640, 2-ethylhexyl 2-ethylhexyphosphonic acid (PC-88A), and tributyl phosphate (TOA), were employed. Cobalt extraction from Cyanex-272, when diluted in Shellsol D70 at 15% volume, achieved optimal results at pH levels between 5.5 and 6. On the other hand, a 10% Acorga M5640 extract of cobalt exhibited an impressive efficiency of over 85% at pH 7, with minimal concurrent lithium extraction. Attempts to strip cobalt from the loaded Acorga phase, even with an increased sulfuric acid concentration, proved unsuccessful. As an alternative, PC-88A and TOA were identified as effective extractants for cobalt. A notable separation factor of nearly 350 was achieved with 90% cobalt extraction at pH 4.5 using 10% PC-88A. A synergistic blend consisting of 10% PC-88A and 5% TOA for cobalt extraction at pH 5.4 demonstrated comparable extraction efficiency while remarkably elevating the separation factor to 1170, four times greater than that of 10% PC-88A in isolation. Consequently, a synergistic mixture of PC-88A and TOA was established as the preferred cobalt extractant for pH levels between 5.5 and 6.0.

Zhao *et al.* (2011) also examined the use of Cyanex-272, PC-88A, and a synergistic extractant mix, both with and without EDTA. When EDTA was added, it made it easier for the water-loving Co-EDTA complex to move into the organic phase. This caused the amount of Co2+ to be more evenly distributed. The concentration of PC-88A was reported as 4.99 mol/L. With 0.03 mol/L Na-Cyanex 272, about 84% of the cobalt and 8% of the lithium were extracted from the organic phase at a pH of 6.9. Choubey *et al.* (2021) and Maroufi *et al.* (2020) separated Co and Li from a sulfate solution comprising 10.44 g/L Co and 1.33 g/L Li using Cyanex 272. With 15% Cyanex 272 and 3% isodecanol, which changed the phase, almost 99.9% of Co was extracted in two steps at pH 5 and an O/A ratio of 1/1. In addition, 8% Li was co-extracted with cobalt and purged with 10% Na_2CO_3 at 1/1 O/A. Raffinate contains pure lithium solution, recoverable as lithium carbonate after cobalt separation. Nguyen *et al.* (2014) used PC-88A to extract cobalt and nickel from a sulfate leach solution that contained 25.1 g/L Co, 2.54 g/L Ni, and 6.2 g/L Li and had a pH of 2.03. Using 60% Na-0.56 mol/L PC-88A in two stages with an O/A phase ratio of 3/1 and an equilibrium pH of 4.5, more than 99% of the Co was obtained.

Almost 99% of Ni could then be extracted at pH 6 using 5% PC-88A. Since it is less soluble at high temperatures, at 90–100°C, the purified lithium solution formed precipitates that enabled lithium recovery as lithium carbonate, as shown in Figure 4 (Eh–pH diagram). Ni and Li were also separated using a diluted sulfate solution (pH 2) that contained 2.54 g/L of Ni and 4.82 g/L of Li. Nguyen *et al.* (2015) used 0.15 mol/L PC-88A in two counter-current phases at O/A: 1/1 and pH 6.5 to obtain more than 99.6% of the nickel. Kang *et al.* (2010) also found that 50% saponified 0.4 mol/L Cyanex 272 at pH 6 equilibrium extracted 98% Co and 1% Ni with a separation factors close to 750s for Co/Li and Co/Ni.

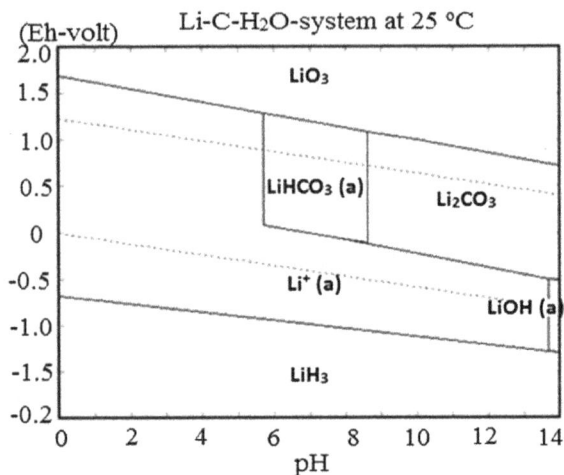

Figure 4.   Eh–pH diagram of lithium in water system (HSC Chemistry 6.0, Li: 1 mol/L). (Adapted from Choubey *et al.*, 2017).

Nickel and lithium co-extraction increased with more than 50% saponified Cyanex-272 for cobalt extraction. After Co and Ni separation, sodium carbonate can precipitate lithium as lithium carbonate at 90°C. Cheng *et al.* (2023) saponified P_5O_7 extracted cobalt from a mother solution of 20 g/L Co, 0.5 g/L Ni, and 2.5 g/L Li. Over 95% of Co was extracted with less than 5% Ni and Li co-extraction using 25% P_5O_7 at 1.5 O/A. Yang *et al.* (2020) used SX to extract Co with 97.8% efficiency using 20 vol. % Mextral-272P at an equilibrium pH of 4.5 and O/A: 1/1 and 15% Li and Ni co-extraction. Finally, Ni and Li are precipitated using attenuated NaOH and Na3PO4 solutions to yield $Ni(OH)_2$ and Li_3PO_4, with 99.13% and 99.67% purities, respectively. According to Yang *et al.* (2020), dimethylglyoxime (DMG) can remove nickel as a Ni–DMG complex.

Next, manganese was extracted at pH 3.5 using 15% di-2-ethylhexyl phosphoric acid (D2EHPA) laden with cobalt and 5% tributyl phosphate (TBP) diluted in kerosene with minimal cobalt co-extraction. Manganese ions remained in the organic phase after cobalt was nearly completely eliminated using diluted oxalic acid. A 0.5 mol/L ammonium oxalate solution $[(NH_4)_2C_2O_4]$ and a heated saturated sodium carbonate solution precipitated cobalt and lithium as cobalt oxalate and lithium carbonate, respectively.

4.2.2 *Chemical precipitation*

Even though solvent extraction may be more effective, several studies have used chemical precipitation to separate Co and Ni by altering the pH. Different precipitating agents separated metals from chloride leach liquid (Wang *et al.*, 2009). Manganese dioxide was precipitated first. When the nickel amine complex $[Ni(NH_3)_6]^{2+}$ was mixed with dimethylglyoxime (DMG) at pH 9, a red nickel-DMG complex formed. This solid complex is separated, redissolved in HCl, and precipitated at a pH of 11 as nickel hydroxide. Nickel is recovered, leaving cobalt hexamine $[Co(NH_3)_6]^{3+}$ in the mother solution. Neutralizing ammonia with hydrochloric acid and settling at a pH of 11 retrieved this. Sodium carbonate was used to extract lithium carbonate at 100°C from a pure lithium solution. Multiple steps and numerous chemicals are needed to separate contaminants in this difficult technique.

Nayl *et al.* (2017) separated metal carbonates using sodium carbonate as a precipitant and sodium hydroxide for pH adjustment. At pH 9.0, 2.0 mol/L of NaOH precipitated around 91% Ni as nickel carbonate ($NiCO_3$). Using NaOH, the pH of the mother solution was increased to 11–12 in order to remove cobalt as cobalt hydroxide. Lithium was precipitated as Li_2CO_3 at 90°C using Na_2CO_3. The precipitate was rinsed with hot water to remove sodium and dried at 100°C for 1 hour to obtain ultrapure lithium carbonate. As discussed in the following, additional economic aspects of the recycling of LIBs also need to be considered.

5. Economic Aspects of Recycling LIBs

The economic aspects of LIB recycling have been examined, considering factors such as operational costs, transportation costs, product utility, and process scale. If recovered materials and avoided disposal costs surpassed collecting and processing expenses, recycling LIBs would be profitable. The recycling procedures of Umicore, Toxco, Inmetco, and Recupyl use pyro-hydro-combined metallurgical methods (Hua *et al.*, 2021). Energy consumption has been accounted for when calculating the process economics of these established processes. As a significant factor, in general, these processes would be nearly identical if they were carried out elsewhere. Consideration has also been given to the metals recovered and their values. As is common knowledge, pyrometallurgical operations are high-calorific processes concentrated solely on recovering cobalt and nickel, leaving lithium and other metals as slag for the cement industry. Hydrometallurgical operations, in contrast, are relatively low-calorific processes that can recover lithium and aluminum in addition to high-value metals. Notably, pyrometallurgy is a versatile process that can accommodate a wide range of LIBs in a single facility (Xu *et al.*, 2022). Thus, the new recycling system incorporates physical, pyro-, and hydro-techniques to reduce waste, decrease energy demand, and maximize metallic value recovery, but it has yet to be commercially tested.

Table 2a compares and contrasts the three operational routes with respect to their energy consumption. The hydrometallurgical process utilizes the least amount of energy and has the capacity to recover lithium. Table 2b presents a comprehensive economic analysis for each phase of the recycling life cycle assessment. The calculation is based on a 5% recycling rate for 2016, which is expected to increase to 10% over the next four years by 2020, resulting in processing 993.5 and 3,974 t of LIBs, respectively. The rate of recovery for the two most essential metals, lithium and cobalt, has been maintained at 95% for both years of calculation. The revenue generated from the recycling of metals is based on their average price in 2016 and their anticipated price in 2020. The results of the analyses in Table 2b's demonstrate unequivocally that the process's profit rises as the rate of recycling and metal price increase. Notably, expenses such as labor and the reclamation of other metals have not been considered, but they might cancel each other out. Consequently, the overall benefit of the process is contingent upon the LIB recycling rate.

Table 2a. Comparative analysis of metallurgical techniques: Pyrometallurgy, hydrometallurgy, and electrometallurgy (Anderson, 2016).

Comparative Factors	Pyrometallurgy	Hydrometallurgy	Electrometallurgy
Capital Expenditure	Elevated	Subdued	Moderate
Energy Consumption	Intermediate	Elevated	Elevated
Dominant Waste Emissions	Solids & Gaseous Effluents	Aqueous & Solid Byproducts	Solids & Gaseous Effluents
Reaction Kinetics	Accelerated	Sluggish	Intermediate
Operational Expenditures	Diminished	Moderate	Elevated
Separation Efficiency	Modest	Enhanced	Modest

Table 2b. Financial evaluation of LIBs Reclamation (Choubey *et al.*, 2017).

Category	2016	2020
A. Cost Breakdown	993.5 (tons) LIBs	3974 (tons) LIBs
1. Operational Expenses per Ton ($2250/ton)	$2,235,375	$8,941,500
2. Transport Costs per Ton ($3039.5)	$3,019,759	$12,078,970
3. Material Handling per Ton ($454/t)	$451,049	$1,804,196
Total Expenditure (Sum of A1, A2, A3)	**$5,706,183**	**$22,824,666**
B. Resource Recovery	230.96 (tons)	943.76 (tons)
1. Cobalt Recovered (tons)	188.76 tons	755 (tons)
2. Lithium Recovered (tons)	42.2 (tons)	188.76 (tons)
C. Total Revenue	$7,786,560	$53,229,920
1. Cobalt Revenue (per ton)	$6,795,360 ($36,000/t)	$49,075,000 ($65,000/t)
2. Lithium Revenue (per ton)	$991,200 ($21,000/t)	$4,154,920 ($22,000/t)
Net Profit (Revenue - Costs)	$2,080,377	$30,405,254

References

Ahamed, A. M., Swoboda, B., Arora, Z., Lansot, J. Y., & Chagnes, A. (2023). Low-carbon footprint diluents in solvent extraction for lithium-ion battery recycling. *RSC Advances*, *13*(33), 23334–23345.

Alavi, N., Adabi, S., Sadani, M., Eslami, A., & Amini, M. M. (2022). Mechanochemical dechlorination of petrochemical sludge through a planetary ball mill and using industrial wastes as additives. *Environmental Progress & Sustainable Energy*, *41*(4), e13828.

Anderson, C. G. (2016). Pyrometallurgy. In *Reference Module in Materials Science and Materials Engineering*, Elsevier, pp. 1–5.

Bowell, R. J., Lagos, L., de los Hoyos, C. R., & Declercq, J. (2020). Classification and characteristics of natural lithium resources. *Elements*, *16*(4), 259–264.

Butt, F. S., Lewis, A., Chen, T., Mazlan, N. A., Wei, X., Hayer, J., Chen, S., Han, J., Yang, Y., Yang, S., & Huang, Y. (2022). Lithium harvesting from the most abundant primary and secondary sources: A comparative study on conventional and membrane technologies. *Membranes*, *12*(4), 373.

Chen, X., Ma, H., Luo, C., & Zhou, T. (2017). Recovery of valuable metals from waste cathode materials of spent lithium-ion batteries using mild phosphoric acid. *Journal of Hazardous Materials*, *326*, 77–86.

Cheng, Q., Marchetti, B., Chen, M., Li, J.-T., Wu, J., Liu, X., & Zhou, X.-D. (2023). Novel approach for in situ recovery of cobalt oxalate from spent lithium-ion batteries using tartaric acid and hydrogen peroxide. *Journal of Material Cycles and Waste Management, 25*(3), 1534–1548.

Choubey, P. K., Chung, K. S., Kim, M. S., Lee, J. C., & Srivastava, R. R. (2017). Advance review on the exploitation of the prominent energy-storage element Lithium. Part II: From sea water and spent lithium ion batteries (LIBs). *Minerals Engineering, 110,* 104–121.

Choubey, P. K., Dinkar, O. S., Panda, R., Kumari, A., Jha, M. K., & Pathak, D. D. (2021). Selective extraction and separation of Li, Co and Mn from leach liquor of discarded lithium ion batteries (LIBs). *Waste Management, 121,* 452–457.

de Oliveira Demarco, J., Cadore, J. S., de Oliveira, Fda. S., Tanabe, E. H., & Bertuol, D. A. (2019). Recovery of metals from spent lithium-ion batteries using organic acids. *Hydrometallurgy, 190,* 105169.

Ding, W., Bao, S., Zhang, Y., Ren, L., Xin, C., Chen, B., Liu, B., & Hou, X. (2023). Stepwise recycling of valuable metals from spent lithium-ion batteries based on in-situ thermal reduction and ultrasonic-assisted water leaching. *Green Chemistry, 25*(17): 6652–6665.

Fu, Y., He, Y., Li, J., Qu, L., Yang, Y., Guo, X., & Xie, W. (2020). Improved hydrometallurgical extraction of valuable metals from spent lithium-ion batteries via a closed-loop process. *Journal of Alloys and Compounds, 847,* 156489.

Garole, D. J., Hossain, R., Garole, V. J., Sahajwalla, V., Nerkar, J., & Dubal, D. P. (2020). Recycle, recover and repurpose strategy of spent Li-ion batteries and catalysts: Current status and future opportunities. *Chem Sus Chem, 13*(12), 3079–3100.

He, Y., Yuan, X., Zhang, G., Wang, H., Zhang, T., Xie, W., & Li, L. (2021). A critical review of current technologies for the liberation of electrode materials from foils in the recycling process of spent lithium-ion batteries. *Science of the Total Environment, 766,* 142382.

Hua, Y., Liu, X., Zhou, S., Huang, Y., Ling, H., & Yang, S. (2021). Toward sustainable reuse of retired lithium-ion batteries from electric vehicles. *Resources, Conservation and Recycling, 168,* 105249.

Ilyas, S., Srivastava, R. R., Singh, V. K., Chi, R., & Kim, H. (2022). Recovery of critical metals from spent Li-ion batteries: Sequential leaching, precipitation, and cobalt-nickel separation using Cyphos IL104. *Waste Management, 154,* 175–186.

JenniLie, J. & ChernLiu, J. (2021). Closed-vessel microwave leaching of valuable metals from spent lithium-ion batteries (LIBs) using dual-function leaching agent: Ascorbic acid. *Separation and Purification Technology, 266,* 118458.

Kamran, U., Heo, Y.-J., Lee, J. W., & Park, S.-J. (2019). Chemically modified activated carbon decorated with MnO2 nanocomposites for improving lithium adsorption and recovery from aqueous media. *Journal of Alloys and Compounds, 794*, 425–434.

Kang, J., Senanayake, G., Sohn, J., & Shin, S. M. (2010). Recovery of cobalt sulfate from spent lithium ion batteries by reductive leaching and solvent extraction with Cyanex 272. *Hydrometallurgy, 100*(3), 168–171.

Li, L., Qu, W., Zhang, X., Lu, J., Chen, R., Wu, F., & Amine, K. (2015). Succinic acid-based leaching system: A sustainable process for recovery of valuable metals from spent Li-ion batteries. *Journal of Power Sources, 282*, 544–551.

Liang, Z., Ding, X., Cai, C., Peng, G., Hu, J., Yang, X., Chen, S., Liu, L., Hou, H., Liang, S., & Xiao, K. (2022). Acetate acid and glucose assisted subcritical reaction for metal recovery from spent lithium ion batteries. *Journal of Cleaner Production, 369*, 133281.

Maroufi, S., Assefi, M., Nekouei, R. K., & Sahajwalla, V. (2020). Recovery of lithium and cobalt from waste lithium-ion batteries through a selective isolation-suspension approach. *Sustainable Materials and Technologies, 23*, e00139.

Meshram, P., Pandey, B. D., & Mankhand, T. R. (2023). Corrigendum to Recovery of valuable metals from cathodic active material of spent lithium-ion batteries: Leaching and kinetic aspects. *Waste Management, 157*, 17.

Mossali, E., Picone, N., Gentilini, L., Rodriguez, O., Pérez, J. M., & Colledani, M. (2020). Lithium-ion batteries towards circular economy: A literature review of opportunities and issues of recycling treatments. *Journal of Environmental Management, 264*, 110500.

Murali, A., Zhang, Z., Free, M. L., & Sarswat, P. K. (2021). A comprehensive review of selected major categories of lithium isotope separation techniques. *Physica Status Solidi (a), 218*(19), 2100340.

Nayl, A. A., Elkhashab, R. A., Badawy, S. M., & El-Khateeb, M. A. (2017). Acid leaching of mixed spent Li-ion batteries. *Arabian Journal of Chemistry, 10*, S3632–S3639.

Nguyen, V. T., Lee, J. C., Jeong, J. K., Kim, B. S., & Pandey, B. D. (2015). The separation and recovery of nickel and lithium from the sulfate leach liquor of spent lithium ion batteries using PC-88A. *Chemical Engineering, 53*(2), 137–144.

Nguyen, V. T., Lee, J., Jeong, J., Kim, B.-S., & Pandey, B. D. (2014). Selective recovery of cobalt, nickel and lithium from sulfate leachate of cathode scrap of Li-ion batteries using liquid-liquid extraction. *Metals and Materials International, 20*, 357–365.

Ning, P., Meng, Q., Dong, P., Duan, J., Xu, M., Lin, Y., & Zhang, Y. (2020). Recycling of cathode material from spent lithium ion batteries using an ultrasound-assisted DL-malic acid leaching system. *Waste Management*, *103*, 52–60.

Ochromowicz, K., Aasly, K., & Kowalczuk, P. B. (2021). Recent advancements in metallurgical processing of marine minerals. *Minerals*, *11*(12), 1437.

Panda, N., Cueva-Sola, A. B., Dzulqornain, A. M., Thenepalli, T., Lee, J-Y., Yoon, H-S., & Jyothi, R. K. (2023). Review on lithium ion battery recycling: challenges and possibilities. *Geosystem Engineering*, 1–18.

Pinegar, H. & Smith, Y. R. (2020). Recycling of end-of-life lithium-ion batteries, Part II: Laboratory-scale research developments in mechanical, thermal, and leaching treatments. *Journal of Sustainable Metallurgy*, *6*, 142–160.

Pourbaix, M. (1966). Atlas of electrochemical equilibria in aqueous solutions. National association of corrosion engineers, Michigan, NACE.

Pranolo, Y., Zhang, W., & Cheng, C. Y. (2010). Recovery of metals from spent lithium-ion battery leach solutions with a mixed solvent extractant system. *Hydrometallurgy*, *102*(1), 37–42.

Srivastava, R. R., Kim, M., & Lee, J. (2013). Separation of tungsten from Mo-rich leach liquor by adsorption onto a typical Fe–Mn cake: Kinetics, equilibrium, mechanism, and thermodynamics studies. *Industrial & Engineering Chemistry Research*, *52*(49), 17591–17597.

Sun, L., Liu, B., Wu, T., Wang, G., Huang, Q., Su, Y., & Wu, F. (2021). Hydrometallurgical recycling of valuable metals from spent lithium-ion batteries by reductive leaching with stannous chloride. *International Journal of Minerals, Metallurgy and Materials*, *28*(6), 991–1000.

Takacova, Z., Orac, D., Klimko, J., & Miskufova, A. (2023). Current Trends in Spent Portable Lithium Battery Recycling. *Materials*, *16*(12), 4264.

Takahashi, V. C. I., Junior, A. B. B., Espinosa, D. C. R., & Tenório, J. A. S. (2020). Enhancing cobalt recovery from Li-ion batteries using grinding treatment prior to the leaching and solvent extraction process. *Journal of Environmental Chemical Engineering*, *8*(3), 103801.

Torkashvand, J., Rezaei Kalantary, R., Heidari, N., Kazemi, Z., Kazemi, Z., Farzadkia, M., Amoohadi, V., & Oshidari, Y. (2021). Application of ultrasound irradiation in landfill leachate treatment. *Environmental Science and Pollution Research*, *28*, 47741–47751.

Waghmare, A., Nagula, K., Pandit, A., & Arya, S. (2019). Hydrodynamic cavitation for energy efficient and scalable process of microalgae cell disruption. *Algal Research*, *40*, 101496.

Wang, D., Zhang, X., Chen, H., & Sun, J. (2018). Separation of Li and Co from the active mass of spent Li-ion batteries by selective sulfating roasting with sodium bisulfate and water leaching. *Minerals Engineering*, *126*, 28–35.

Wang, K., Hu, T., Shi, P., Min, Y., Wu, J., & Xu, Q. (2021). Efficient recovery of value metals from spent lithium-ion batteries by combining deep eutectic solvents and coextraction. *ACS Sustainable Chemistry & Engineering, 10*(3), 1149–1159.

Wang, S., Tian, Y., Zhang, X., Yang, B., Wang, F., Xu, B., Liang, D., & Wang, L. (2020). A Review of Processes and Technologies for the Recycling of Spent Lithium-ion Batteries. *IOP Conference Series: Materials Science and Engineering, 782*(2), 22025.

Wang, Y., Xu, Z., Zhang, X., Yang, E., & Tu, Y. (2022). A green process to recover valuable metals from the spent ternary lithium-ion batteries. *Separation and Purification Technology, 299*, 121782.

Wen, J. & Lee, M. S. (2023). Recovery of nickel and cobalt metal powders from the leaching solution of spent lithium-ion battery by solvent extraction and chemical reduction. *Mineral Processing and Extractive Metallurgy Review*, 1–11.

Xu, C., Steubing, B., Hu, M., Harpprecht, C., van der Meide, M., & Tukker, A. (2022). Future greenhouse gas emissions of automotive lithium-ion battery cell production. *Resources, Conservation and Recycling, 187*, 106606.

Xu, M., Kang, S., Jiang, F., Yan, X., Zhu, Z., Zhao, Q., Teng, Y., & Wang, Y. (2021). A process of leaching recovery for cobalt and lithium from spent lithium-ion batteries by citric acid and salicylic acid. *RSC Advances, 11*(44), 27689–27700.

Yan, S., Sun, C., Zhou, T., Gao, R., & Xie, H. (2021). Ultrasonic-assisted leaching of valuable metals from spent lithium-ion batteries using organic additives. *Separation and Purification Technology, 257*, 117930.

Yang, Y., Song, S., Lei, S., Sun, W., Hou, H., Jiang, F., Ji, X., Zhao, W., & Hu, Y. (2019). A process for combination of recycling lithium and regenerating graphite from spent lithium-ion battery. *Waste Management, 85*, 529–537.

Yang, Y., Liu, F., Song, S., Tang, H., Ding, S., Sun, W., Lei, S., & Xu, S. (2020). Recovering valuable metals from the leaching liquor of blended cathode material of spent lithium-ion battery. *Journal of Environmental Chemical Engineering, 8*(5), 104358.

Yazdani, E., Hashemabadi, S. H., & Taghizadeh, A. (2019). Study of waste tire pyrolysis in a rotary kiln reactor in a wide range of pyrolysis temperature. *Waste* Management, *85*, 195–201.

Zhang, G., Du, Z., He, Y., Wang, H., Xie, W., & Zhang, T. (2019). A sustainable process for the recovery of anode and cathode materials derived from spent lithium-ion batteries. *Sustainability, 11*(8), 2363.

Zhao, J., Zhang, B., Xie, H., Qu, J., Qu, X., Xing, P., & Yin, H. (2020). Hydrometallurgical recovery of spent cobalt-based lithium-ion battery cathodes using ethanol as the reducing agent. *Environmental Research, 181*, 108803.

Zheng, X., Zhu, Z., Lin, X., Zhang, Y., He, Y., Cao, H., & Sun, Z. (2018). A mini-review on metal recycling from spent lithium ion batteries. *Engineering, 4*(3), 361–370.

Zhou, F., Li, X., Wang, S., Qu, X., Zhao, J., Wang, D., Chen, Z., & Yin, H. (2023). Recovery of valuable metals from spent lithium-ion batteries through biomass pyrolysis gas-induced reduction. *Journal of Hazardous Materials, 459,* 132150.

Zhuang, L., Sun, C., Zhou, T., Li, H., & Dai, A. (2019). Recovery of valuable metals from LiNi0. 5Co0. 2Mn0. 3O$_2$ cathode materials of spent Li-ion batteries using mild mixed acid as leachant. *Waste Management, 85,* 175–185.

Chapter 4

Recycling Technologies and Resource Recovery from Plastic Waste

K. Hephzi Jones and Pankaj Pathak*

*Resource Management Lab, Department of Environmental Science
and Engineering, SRM University - AP, Mangalagiri,
Andhra Pradesh-522240, India*
**pankajpathak18@gmail.com*

Abstract

Plastics have become one of the most integral parts of our day-to-day lives; however, after use, plastic waste accumulates on land and in water bodies, which wreaks havoc in the environment by releasing toxic gases. As the usage of plastic increases, it will result in the depletion of natural resources and greater difficulties in managing plastic waste. Accordingly, in this chapter, we focus on various types of resources, such as fuel, gas, energy, and electricity, which are recovered from plastic waste through different waste management technologies such as primary, secondary, tertiary, and quaternary recycling. This chapter is set forth by critically evaluating different types of plastic waste technologies, the products, and their byproducts formed during tertiary treatment. Material and resource recovery of different types of waste plastics through gasification and pyrolysis offers significant environmental benefits by promoting resource extraction and reducing the associated environmental impacts linked to the extraction process.

61

Keywords: Plastic waste, gasification, pyrolysis, incineration, methanolysis, hydrogenation.

1. Introduction

The term "pliable," which means "easily shaped," was the ancestor of the word "plastic." Plastics are also referred to as polymers or "long chains of monomers," which are joined by additional identical subunits to form polymers. Plastics can easily be transformed from one shape to another according to the needed functionality (Evode *et al.*, 2021; Geyer *et al.*, 2017). The production of Bakelite in 1907 followed that of the first plastic material, which was produced in 1850 (Horton, 2022; Liu *et al.*, 2022). The 1950s and 1960s saw the actual advent of new plastics and their products. Plastics have undergone a dramatic transformation in just over a century, from being acclaimed as a scientific marvel to being despised as an environmental disaster (Letcher, 2020; Thompson *et al.*, 2009).

Based on their physical and chemical characteristics, several kinds of plastics can be distinguished, including thermoplastics and thermosets (Qureshi *et al.*, 2020; Kibria *et al.*, 2023). Thermoplastics can be heated, melted, and molded, then cooled to become hard. Due to their plastic nature, they are also mechanically recyclable, which is an excellent waste management strategy (Iacovidou *et al.*, 2019; Maqsood & Altaf, 2023). Thermosetting, or thermoset, plastics are artificial materials that undergo several physicochemical transformation processes through various heat treatments to form a three-dimensional network. It is not possible to undo this alteration (Evode *et al.*, 2021).

Every year 0.4 billion tons of plastic waste are produced globally, out of which 9% is recycled, 12% is incinerated, and the remaining is landfilled or open dumped. Packaging uses over 40% of the plastics produced globally (Kibria *et al.*, 2023; Tejaswini *et al.*, 2022).

The most well-known plastics include polyvinyl chloride (PVC) (cable insulations, pipes, multilayer tubes, etc.), polystyrene (PS) (disposable cups, foams, CD covers, etc.), polyethylene terephthalate (PET) (water bottles, plastic films, sheets, etc.), polyethylene (PE), low-density polyethylene (LDPE) (bottles, plastic toys, milk pouches, etc.), high-density polyethylene (HDPE) (shopping bags, metalized pouches, detergent bags, etc.), linear low-density polyethylene (LLDPE), polypropylene (PP) (bottle caps, straws, medicine bottles, etc.), and

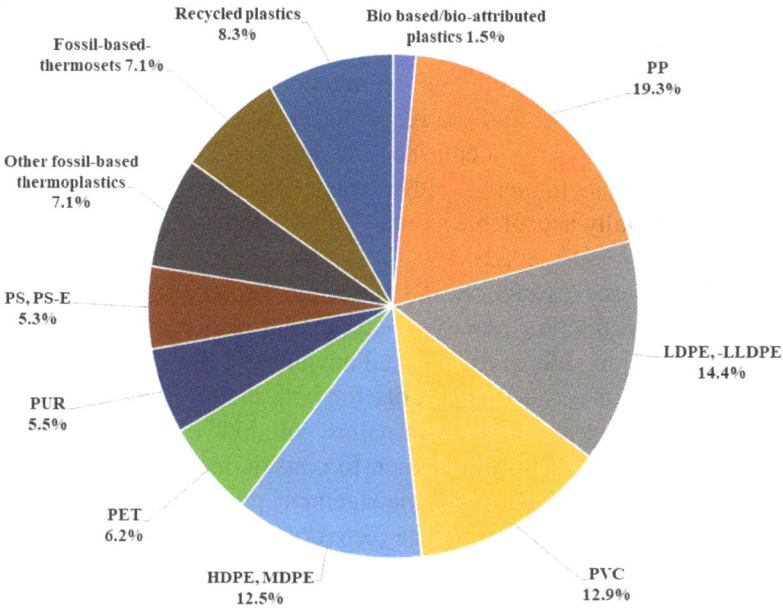

Figure 1.   Demand percentage for different types of plastic.
Source: Plastics Europe, 2022.

Figure 2.   Society of Plastics Industry packaging codes.

polytetrafluoroethylene (PTFE, or Teflon), refer Figures 1 and 2. Plastics have revolutionized living standards by replacing metals, wood, and cement due to their lightweight and excellent durability (Karmakar, 2022).

Plastic wastes made of thermoset, thermoplastic, and elastomers can significantly contaminate the environment since they are difficult to break down (Huang *et al.*, 2022). The solution to many environmental and sustainability problems is therefore proper plastic waste management. According to the "take-make-consume-waste" method, the conventional model is based on the exploitation of resources. A circular-economy-based, generative, and restorative paradigm must be adopted in place of the linear approach (Singh & Ordoñez, 2016). Plastic waste is typically managed via methods such as landfilling, mechanical pulverization, incineration, recycling, microbiological degradation, and thermal breakdown. The most prominent method for recycling energy from plastic waste is incineration since it allows a considerable amount of energy to be generated and utilized in numerous other ways. The most environmentally friendly and socially acceptable solution out of all those is recycling plastic waste (Naderi *et al.*, 2023).

As for resource extraction and reducing the related environmental impacts of the extraction process, material and resource recovery from waste offer considerable environmental advantages. Simply put, recovering resources from waste has benefits for both the environment and the economy. Obtaining resources from nature could have negative environmental effects. As a result, in addition to considering the environmental benefits, the evaluation of waste management performance should also consider the full benefits of resource recovery.

2. Different Technologies for Plastic Waste Treatment

According to ASTM Standard D5033, there are four stages in the plastic recycling process: primary, secondary, tertiary, and quaternary, refer Figure 3. Recycling of plastic trash can be divided into chemical, mechanical, and biological categories based on the mechanisms used in each process (Naderi *et al.*, 2023).

2.1 *Primary recycling*

In a closed-loop system, processes are designed to maximize resource efficiency, recycle resources, and decrease the discharge of harmful gases to minimize waste and lessen the environmental impact (Maqsood & Altaf, 2023). Uncontaminated plastic is mechanically reprocessed, producing a resin or pellets of a similar grade. Plastic that is multi-layered or mixed with contaminants cannot be recycled; only homogeneous or well-sorted plastic may be recycled in the primary recycling stage (Iacovidou *et al.*, 2019). The utilization of PET recovered from post-consumer bottles in the manufacture of new bottles is an illustration of primary recycling (Merrington, 2017).

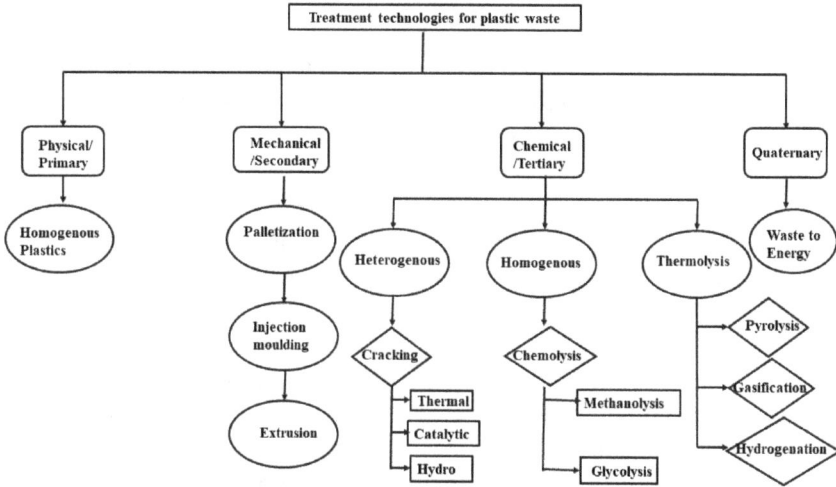

Figure 3.   Different treatment technologies for Plastic waste.

2.2 *Secondary recycling*

Treatment of plastic solid waste (PSW) involves the use of mechanical recycling on a larger scale. The recycling process involves washing away organic residue, followed by shredding, melting, and remolding of the polymer to impart specific qualities for manufacturing (Ribeiro *et al.*, 2016). The recovered plastic is utilized in goods with less stringent performance specifications than the original application. To meet the requirements of the new product, secondary recycling frequently necessitates reformulation. Making floor tiles from a mixture of polyolefins is an example of secondary recycling (Merrington, 2017).

2.3 *Tertiary recycling*

Tertiary plastic recycling is distinguished by depolymerization into oligomers and monomers through a targeted biochemical approach. These methods include hydrolysis or methanolysis, or rupturing the hydrocarbon backbone, such as through gasification or pyrolysis, to break the material into its component fragments (Lee & Liew, 2021).

2.3.1 *Pyrolysis*

One of the best processes for converting mass to energy (E) with gaseous and liquid products of high energy values is pyrolysis, which is a proper approach for recovering energy from waste plastics. Thermal cracking, or pyrolysis, is the thermal breakdown of large-polymer chain molecules into simpler small molecules. The process takes place briefly in the absence of oxygen under elevated pressure and temperature. Up to 80% (wt.) of liquid oil can be produced through pyrolysis at temperatures around 500–800°C (Dai *et al.*, 2022; Dargo Beyene, 2014). To create products tailored to individual preferences, the process parameters can be changed. Consequently, pyrolysis is frequently described as a flexible process. The liquid oil generated is of good quality since it can be applied to several applications without needing further upgrades or treatment. Because the gaseous fuel, which is produced as a byproduct of pyrolysis, has a good calorific value, it can be used to offset the energy needs of the pyrolysis plant. Since pyrolysis is easier to handle and more adaptable than traditional recycling methods, it is more frequently used. Additionally, pyrolysis requires less intensive sorting of waste compared to other processes, making it less labor-intensive (Qureshi *et al.*, 2020; Vijayakumar & Sebastian, 2018).

2.3.2 *Gasification*

Gasification is the process of converting the chemical energy contained in solid fuel into gaseous fuel. Syngas containing carbon monoxide (CO) and hydrogen (H) is produced by the conversion of solid fuel. Gasifier reactions, on the other hand, are quite complex (Lopez *et al.*, 2018). Because of this, supplementary products are produced depending on the characteristics of the solid fuel and processing conditions, such as the relative contents of CO_2, CH_4, tar, and HCl. To prevent combustion, the fuel is reacted at high temperatures of 1000–1500°C in a low-oxygen environment (Tejaswini *et al.*, 2022). Gasification serves as a platform technology for the majority of petrochemical processes. It provides a pathway for power generation in fuel cells and turbines, the production of chemical synthesis and liquid fuels, and other applications (Saebea *et al.*, 2020; Salaudeen *et al.*, 2018). The configurations of gasifiers are categorized as fixed/moving bed, fluidized bed, and spouted bed gasifiers depending on how the gas and fuel come into contact there (Mishra & Upadhyay, 2021).

2.3.3 *Hydrogenation*

Plastics can be hydrogenated as an alternative to breaking down the chain polymer. When compared to treatments without hydrogen, hydrogenation produces extremely saturated products, which prevents the presence of olefins in the liquid fraction and favors their use as fuels without additional processing. The elimination of heteroatoms such as nitrogen (N), chlorine (Cl), and sulfur (S) in the form of volatile molecules is also facilitated by hydrogenation. However, hydrogenation has several limitations, mainly because hydrogen is expensive and it requires operation at high pressure. Hydrochloric acid, halogenated solid waste, and gas are the primary byproducts of hydrogenating solid waste made of plastic (Awasthi *et al.*, 2017; Dargo Beyene, 2014).

2.3.4 *Methanolysis*

Ethylene glycol (EG) and dimethyl terephthalate (DMT) are the major products of methanolysis, which involves the breakdown of polyethylene terephthalate (PET) by methanol at high pressures and temperatures. PET flakes are typically methanolized at temperatures ranging between 180°C and 280°C and at pressures between 2 and 4 MPa. Currently, plant waste, discarded films, fiber waste, and scrap bottles are all successfully treated by methanolysis (Crippa & Morico, 2019; Al-Sabagh *et al.*, 2016).

2.3.5 *Glycolysis*

Glycolysis is the commonly used technology for the chemical recycling of PET waste, in which plastic polymers are decomposed into their constituent monomers through reactions with glycols such as EG in the presence of a catalyst. The main product of deep glycolysis in EG is bis-2-(hydroxyethyl) terephthalate (BHET), which can be used directly to make PET. The process is carried out at temperatures ranging between 180 and 250°C with excessive EG and in the presence of a transesterification catalyst, zinc, or lithium acetate (Thachnatharen *et al.*, 2021; Al-Sabagh *et al.*, 2016).

2.4 *Quaternary recycling*

The most popular thermal conversion technology currently in use is incineration, which converts waste into energy as heat and electricity by

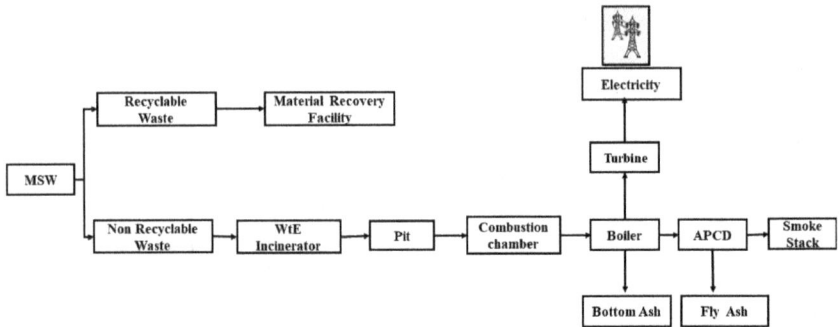

Figure 4.   Process of municipal solid waste incineration.

carefully regulating the combustion of MSW at high temperatures (850–1200°C), refer Figure 4 (Karmakar *et al.*, 2023; Kumar, 2020). Carcinogenic volatile organic chemicals, particulate matter, dioxins, and furans are released during the energy recovery process from plastic waste. These substances degrade chlorofluorocarbon (CFC) and other blowing agents, posing a serious risk to human health and the environment (Tejaswini *et al.*, 2022).

3. Resource Recovery from Plastic Waste

Currently, primary (1°), secondary (2°), tertiary (3°), and quaternary (4°) plastic solid waste recycling techniques have been developed, as seen in Figure 3 (Singh *et al.*, 2017). However, these processes can only be used with clean, single-use plastics and result in polymers with inferior qualities compared to those of virgin materials. Currently, primary and secondary recycling are commonly used due to their simplicity of use and the potential to yield valuable goods. The quality of the products produced through mechanical recycling, which now dominates recycling practices, is also impacted by severe contamination and a wide diversity of plastics in the waste stream. Beyond mechanical recycling, current plastic solid waste recycling heavily relies on "quaternary" (4°) recycling, which contaminates neighboring landscape surfaces and emits polluting components, including volatile organic compounds (VOCs), SO_x, NO_x, polychlorinated dibenzo-furans (PCDFs), and dioxins (Singh *et al.*, 2017; Singh & Ordoñez, 2016).

"Tertiary" (3°) recycling, which is more effective, focuses on depolymerizing plastic waste to create fuels and chemicals that could replace petroleum-based goods (Dai *et al.*, 2022). Because it can convert plastic waste into extremely valuable fuels and chemicals while being environmentally acceptable, pyrolysis has recently attracted increased attention among the available thermochemical methods. Pyrolysis, in contrast to gasification and incineration, is carried out in the absence of oxygen to reduce harmful contaminants and CO_2 emissions. Particularly, pyrolytic conditions (lower than 600°C and the absence of oxygen) can significantly decrease polychlorinated dibenzofurans (PCDF) and polychlorinated dibenzo-para-dioxins (PCDD). Additionally, pyrolysis offers distinct benefits in terms of product value, operating costs, and the generation of alternative energy (Singh *et al.*, 2019).

3.1 Resource recovery of plastics from pyrolysis

At temperatures exceeding 400°C, pyrolysis, a flexible thermal breaking process, occurs in the absence of oxygen. Fast, slow, and flash are the three types of processes in pyrolysis (Papari *et al.*, 2021). Pure nitrogen is utilized in pyrolysis to provide the necessary neutral environment and facilitate product removal from the reactors. The pyrolysis product spectrum is strongly influenced by the reactor selection. The composition of the feedstock has a major impact on the value of pyrolysis-derived products, but reactor types also play a significant role in maximizing the intended product (Qureshi *et al.*, 2020). There are two types of pyrolysis: thermocatalytic and purely thermal. Mostly gaseous materials are produced during thermal pyrolysis. Up to 93% of the total recovered products — which include a wide range of components — are gaseous byproducts of thermal pyrolysis. At temperatures in the range of 350–600°C, thermo-catalytic pyrolysis speeds up the degradation of polymers. In catalytic pyrolysis, many catalysts are used, including zeolite-based catalysts (ZSM-5), CoMo oxide/Al_2O_3, and ZnO. By using catalysts, one can improve liquid recovery while reducing char and ash levels (Pandey *et al.*, 2020; Vijayakumar & Sebastian, 2018).

3.1.1 Types of Pyrolysis

Through chemical and thermal reactions, the thermochemical process breaks down the long-chain polymer molecules into smaller and less complex ones. Slow pyrolysis usually occurs between 350 and 550°C, with the rate of heating ranging from 1 to 10°C/minute and a lengthy residence time for the vapor. Char is the main byproduct of slow pyrolysis due to the slow heating rate favoring the formation of solids among multiple collateral-competitive reactions (Papari *et al.*, 2021). In most cases, fast pyrolysis occurs between 500 and 700°C. The feedstock is heated at a rate of over 1000°C/minute, with vapor residence times typically lasting only a few seconds. In flash pyrolysis, the temperature is typically above 700°C, the feed is heated up very quickly, and the vapor residence times are in the millisecond range. For biomass feedstocks, flash pyrolysis can produce more oil than fast pyrolysis; however, plastic waste behaves differently since it produces more gas than the other products. Plastic wastes of all types, either alone or in combinations, are pyrolyzed to produce liquids/waxes, solid residues, and gases as byproducts (Papari *et al.*, 2021; Papari & Hawboldt, 2017).

Plastic oil produced could be in wax or liquid form. Wax is primarily composed of alkenes and alkane hydrocarbons with high boiling points (C20+) and high viscosity at room temperature. Wax is often an intermediate product that needs to go through an additional process, such as fluid catalytic cracking (FCC), to be transformed into a liquid fuel. Mono- and polyaromatic chemicals, as well as aliphatic compounds, make up the majority of liquid plastic oil. Plastic oil can serve as a precursor for fuel applications as well as an intermediate that can be further cracked at higher temperatures and extremely short contact durations to produce ethylene and propylene (Pandey *et al.*, 2020). The "gas product" consists mainly of ethylene, methane, propylene, ethane, butane, and butadiene. The process can be made self-sustaining and independent of external energy sources by using the gas byproduct as a source of pyrolysis energy. In addition, it is possible to extract and collect valuable olefin components from the gas stream for chemical recycling. The residual pyrolysis product, known as the solid residue, is primarily composed of ash and coke (Papari *et al.*, 2021).

3.1.2 *Process of Pyrolysis to produce fuel from various types of plastics*

The pyrolysis technique has proven to be a potential method for using waste plastic to generate electricity. Instead of letting the waste pile up in landfills, through this technique, valuable liquid oil, gaseous fuel, and char can be produced from it, thus greatly contributing to mitigating environmental damage refer Figure 5. In terms of fuel handling and storage,

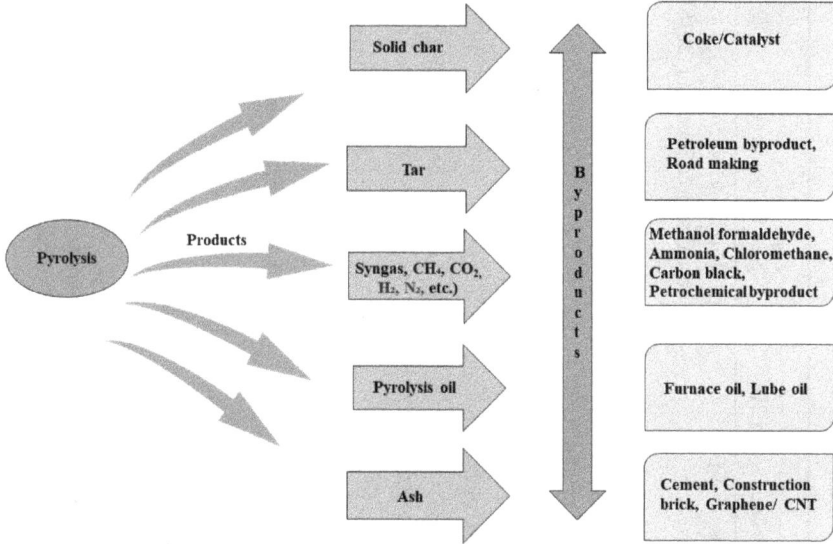

Figure 5.   Products and by products in the Plastic waste through Pyrolysis.

liquid fuel is preferred for small-scale batch pyrolysis processes. In contrast to PET, the PS plastic category has been determined to have the highest liquid yield refer Table 1. Pyrolysis is not recommended for the PVC category because it produces highly hazardous HCl gas. However, pyrolysis is a dependable and long-term approach to addressing the nation's mounting plastic waste problem, which is getting worse by the day (Singh *et al.*, 2019; Vijayakumar & Sebastian, 2018).

3.1.3 *Production of graphene nanosheet through Pyrolysis from waste plastic*

PP, PET, and PE are the most abundant plastics in the environment because of their high carbon content and high annual demand. After segregation, the shredding machine and washing unit of a pilot-scale plant are used to shred and thoroughly wash the materials, respectively. In order to retain the product's uniformity after this stage, washed and mixed polymers are appropriately combined with Al_2O_3, which can be utilized as a degradant for the formation of a carbon skeleton during the pyrolysis process. Following the pyrolysis secondary stage at 750°C, the Al_2O_3-mixed polymers are fed into

Table 1.　Production of fuel from various types of plastics through pyrolysis.

Types of plastic	Type of reactor	Temperature	Time/Rate of Heat (°C/min)	Fluidizing medium	Products (wt %)	References
PET	Fixed bed reactor	500°C	10	Nitrogen	Gas (76.9) oil (23.1)	(Çepelioğullar & Pütün, 2013; Vijayakumar & Sebastian, 2018)
PET	Fixed bed reactor	500°C	10	Nitrogen	Gas (52.13) Oil (39.89) Solid residue (8.98)	(Fakhrhoseini & Dastanian, 2013; Vijayakumar & Sebastian, 2018)
HDPE	Steel reactor	300–350°C		Nitrogen	Gas (17.24) Oil (80.88) Solid (1.88)	(Shah et al., 2023; Vijayakumar & Sebastian, 2018)
PVC	Vacuum batch	520°C	10		Liquid (0.45–12.79) HCl (58.2)	(Vijayakumar & Sebastian, 2018)
LDPE	Fixed bed	500°C	20	Nitrogen	Gas (5) Oil (95)	(Vijayakumar & Sebastian, 2018)
PP	Horizontal steel	300°C	30		Gas (28.84) Oil (69.82) Solid residue (1.34)	(Vijayakumar & Sebastian, 2018)
PS	Semi batch	400°C	10		Gas (6) Oil (90) Solid residue (4)	(Vijayakumar & Sebastian, 2018)
Mixed plastic	Fluidized bed	650°C			Oil (48)	(Donaj et al., 2012; Vijayakumar & Sebastian, 2018)

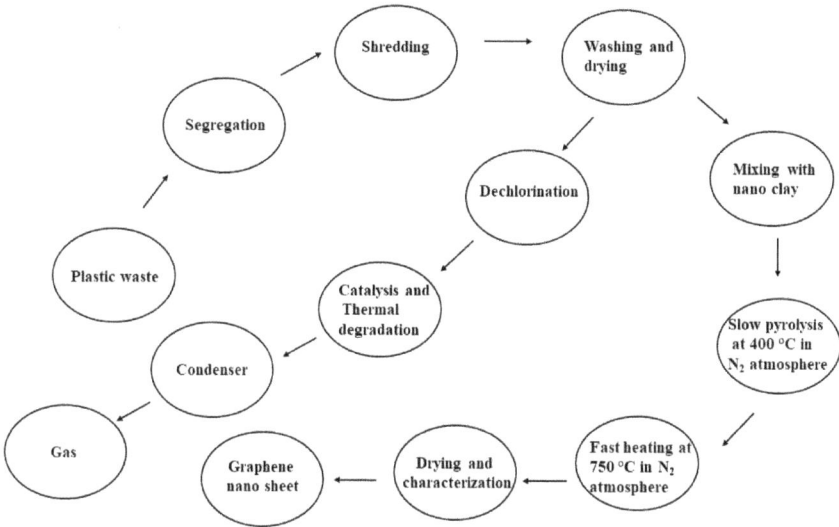

Figure 6.   Process of graphene nanosheet and gas from waste plastic.

the primary reactor (1°) constructed of nickel (Ni) metal, which also serves as a catalytic bed, at a temperature of 400°C. In order to achieve reduced pure graphene nanosheets (GNs), the resulting reduced graphene nanosheets are transported to a ball mill, which produces extremely small particles of the reduced graphene nanosheets. These particles are again washed with 5% HCl and repeatedly rinsed with distilled water refer Figure 6 (Garg *et al.*, 2022; Pandey *et al.*, 2019; Zhao *et al.*, 2022).

3.1.4 *Resource recovery of plastics through gasification*

In the process of gasification, a mixture of CO and H_2, which forms a fuel gas called syngas, is produced. When carbonaceous waste reacts with a gasification agent (oxygen and/or steam) at temperatures ranging from 700 to 1600°C under partially oxidizing conditions (i.e., the absence or sub-stoichiometric presence of oxygen). To increase the thermodynamic efficiency of the downstream power cycle, gasification allows syngas to be burned at higher temperatures than those achievable with the original fuel or even in fuel cells. Additionally, problematic chemical components such as chloride and potassium can be removed from syngas to produce clean combustion flue gases (Article Author *et al.*, 2019). The most

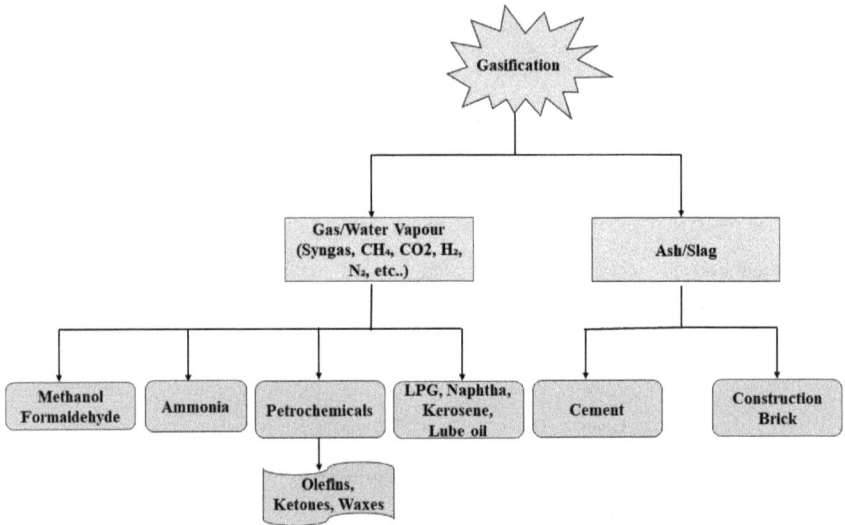

Figure 7.　Formation of products and by products in gasification of plastic waste.

undesirable byproducts of plastic gasification, char and tar, are intended to be converted to syngas or other gaseous products at the maximum possible rate refer Figure 7. Gasification is a complicated process that involves numerous chemical reactions. The main four steps in the gasification process are (1) drying, (2) pyrolysis, (3) oxidation (O), and (4) reduction. There are different types of chemical reactors for the gasification of plastic waste, including fixed (packed) bed reactors, spouted conical bed gasifiers, fluidized bed reactors, and plasma gasification reactors refer Table 2 (Pandey *et al.*, 2019).

The primary issue with gasifying waste plastic is the production of tar, which creates significant operational difficulties, lowering the gas yield and affecting the productivity of the entire process.

4. Conclusion

Even though plastics play a vital role in our daily lives, due to their physiochemical composition, if post-use disposal is not properly handled, they can harm both human life and the environment. The recycling and regeneration of plastic waste could reduce the demand for feedstock made of non-renewable fossil fuels as well as the pollution it causes to the

Table 2. Various compositions in steam plastic waste gasification obtained by different authors (Erkiaga *et al.*, 2013; Lopez *et al.*, 2016; Shah *et al.*, 2023).

Type of Plastic	Reactor	Reaction conditions	Gas composition (% vol)	Gas produced (m^3/kg)	LHV (MJ/m^3)	Tar content (g/m^3)
Waste plastics	Plasma reactor	Gasifying agent: steam/O_2 T: 1200	CO:34, H_2:62, CH_4; CO_2	3.5	10.1	—
PE	Spouted bed reactor	T:900, S/P:1	CO:27, H_2:58, CH_4: 7, CO_2:3	3.2	16.2	15
PET	Semi-batch and fixed bed reactor	T:1000	CO:6, H_2:61, CH_4: 2, CO_2:12	—	7.8	—
PP	Fluidized bed (dual)	T:850, S/P:2	CO:4, H_2:34, CH_4: 40, CO_2:8	1	27.2	180
HDPE	Fixed (packed) batch bed	Gasifying agent: steam/O_2 1:1, T:850	CO:43, H_2:35, CH_4: 11, CO_2:10	24	—	17
PS	Fixed (packed) batch bed	Gasifying agent: steam/O_2 1:1, T:850	CO:43, H_2:29, CH_4: 1.7, CO_2:26	1.3	—	290

environment. With improved methods of waste collection, segregation, and processing, the recycling and recovery of waste plastics have recently increased globally. Shortly, plastic regeneration and recovery might create a desirable industry that produces high-quality plastic products. This would be possible with international quality certification and an open monitoring mechanism.

References

Al-Sabagh, A. M., Yehia, F. Z., Eshaq, G., Rabie, A. M., & ElMetwally, A. E. (2016). Greener routes for recycling of polyethylene terephthalate. *Egyptian Journal of Petroleum*, *25*(1), 53–64. https://doi.org/10.1016/j.ejpe.2015. 03.001

Article Author, J., Acker, V., Passel, V., Carlos Hernández Parrodi, J., Lucas, H., Gigantino, M., Sauve, G., Laurence Esguerra, J., Einhäupl, P., Vollprecht, D., Pomberger, R., Friedrich, B., Van Acker, K., Krook, J., Svensson, N., & Van Passel, S. (2019). *Integration of resource recovery into current waste management through (Enhanced) landfill mining Rights/ license: Creative Commons Attribution-NonCommercial-NoDerivatives 4.0 International-NC-ND license INTEGRATION OF RESOURCE RECOVERY INTO CURRENT WASTE MANAGEMENT THROUGH (ENHANCED) LANDFILL MINING Keywords: Landfill mining strategies Enhanced landfill mining Resource recovery Waste management practices and policies Economic assessment Environmental impacts*. https://doi.org/10.3929/ethz-b-000426919.

Awasthi, A. K., Shivashankar, M., & Majumder, S. (2017). Plastic solid waste utilization technologies: A Review. *IOP Conference Series: Materials Science and Engineering*, *263*(2), 022024. https://doi.org/10.1088/1757-899X/263/2/022024.

Çepelioğullar, Ö. & Pütün, A. E. (2013). Thermal and kinetic behaviors of biomass and plastic wastes in co-pyrolysis. *Energy Conversion and Management*, *75*, 263–270. https://doi.org/10.1016/j.enconman.2013.06.036.

Crippa, M. & Morico, B. (2019). PET depolymerization: A novel process for plastic waste chemical recycling. *Studies in Surface Science and Catalysis*, *179*, 215–229. https://doi.org/10.1016/B978-0-444-64337-7.00012-4.

Dai, L., Zhou, N., Lv, Y., Cheng, Y., Wang, Y., Liu, Y., Cobb, K., Chen, P., Lei, H., & Ruan, R. (2022). Pyrolysis technology for plastic waste recycling: A state-of-the-art review. In *Progress in Energy and Combustion Science* (Vol. 93). Elsevier Ltd. https://doi.org/10.1016/j.pecs.2022.101021.

Dargo, B. H. (2014). Recycling of plastic waste into fuels, a review. *International Journal of Science, Technology and Society*, *2*(6), 190. https://doi.org/10.11648/j.ijsts.20140206.15.

Donaj, P. J., Kaminsky, W., Buzeto, F., & Yang, W. (2012). Pyrolysis of polyolefins for increasing the yield of monomers' recovery. *Waste Management,* *32*(5), 840–846. https://doi.org/10.1016/j.wasman.2011.10.009.

Erkiaga, A., Lopez, G., Amutio, M., Bilbao, J., & Olazar, M. (2013). Syngas from steam gasification of polyethylene in a conical spouted bed reactor. *Fuel,* *109*, 461–469. https://doi.org/10.1016/j.fuel.2013.03.022.

Evode, N., Qamar, S. A., Bilal, M., Barceló, D., & Iqbal, H. M. N. (2021). Plastic waste and its management strategies for environmental sustainability. *Case Studies in Chemical and Environmental Engineering, 4*, 100142. https://doi. org/10.1016/j.cscee.2021.100142.

Fakhrhoseini, S. M., & Dastanian, M. (2013). Predicting pyrolysis products of PE, PP, and PET using NRTL activity coefficient model. *Journal of Chemistry.* https://doi.org/10.1155/2013/487676.

Garg, K. K., Pandey, S., Kumar, A., Rana, A., Sahoo, N. G., & Singh, R. K. (2022). Graphene nanosheets derived from waste plastic for cost-effective thermoelectric applications. *Results in Materials, 13*, 100260. https://doi. org/10.1016/j.rinma.2022.100260.

Geyer, R., Jambeck, J. R., & Law, K. L. (2017). Production, use, and fate of all plastics ever made. *Science Advances, 3*(7), e1700782. https://doi.org/10.1126/ sciadv.1700782.

Horton, A. A. (2022). Plastic pollution: When do we know enough? *Journal of Hazardous Materials, 422*, 126885. https://doi.org/10.1016/j.jhazmat.2021. 126885.

Huang, S., Wang, H., Ahmad, W., Ahmad, A., Vatin, N. I., Mohamed, A. M., Deifalla, A. F., & Mehmood, I. (2022). Plastic waste management strategies and their environmental aspects: A scientometric analysis and comprehensive review. *International Journal of Environmental Research and Public Health, 19*(8), 4556. https://doi.org/10.3390/ijerph19084556.

Iacovidou, E., Velenturf, A. P. M., & Purnell, P. (2019). Quality of resources: A typology for supporting transitions towards resource efficiency using the single-use plastic bottle as an example. *Science of the Total Environment, 647*, 441–448. https://doi.org/10.1016/j.scitotenv.2018.07.344.

Karmakar, A., Daftari, T., Sivagami, K., Chandan, M. R., Shaik, A. H., Kiran, B., & Chakraborty, S. (2023). A comprehensive insight into waste to energy conversion strategies in India and its associated air pollution hazard. *Environmental Technology and Innovation, 29*, 103017. https://doi.org/ 10.1016/j.eti.2023.103017.

Karmakar, G. P. (2022). Regeneration and recovery of plastics. In *Encyclopedia of Materials: Plastics and Polymers* (Vols. 1–4, pp. 634–651). Elsevier. https://doi.org/10.1016/B978-0-12-820352-1.00045-6.

Kumar, R. (2020). Tertiary and quaternary recycling of thermoplastics by additive manufacturing approach for thermal sustainability. *Materials Today:*

Proceedings, *37*(Part 2), 2382–2386. https://doi.org/10.1016/j.matpr. 2020.08.183.

Lee, A. & Liew, M. S. (2021). Tertiary recycling of plastics waste: an analysis of feedstock, chemical and biological degradation methods. *Journal of Material Cycles and Waste Management*, *23*(1), 32–43. https://doi.org/10.1007/ s10163-020-01106-2.

Letcher, T. M. (2020). Introduction to plastic waste and recycling. In *Plastic Waste and Recycling: Environmental Impact, Societal Issues, Prevention, and Solutions* (pp. 3–12). Elsevier. https://doi.org/10.1016/B978-0-12-817880-5.00001-3.

Liu, X., Lei, T., Boré, A., Lou, Z., Abdouraman, B., & Ma, W. (2022). Evolution of global plastic waste trade flows from 2000 to 2020 and its predicted trade sinks in 2030. *Journal of Cleaner Production*, *376*, 134373. https://doi.org/ 10.1016/j.jclepro.2022.134373.

Lopez, G., Alvarez, J., Amutio, M., Arregi, A., Bilbao, J., & Olazar, M. (2016). Assessment of steam gasification kinetics of the char from lignocellulosic biomass in a conical spouted bed reactor. *Energy*, *107*, 493–501. https://doi.org/10.1016/j.energy.2016.04.040.

Lopez, G., Artetxe, M., Amutio, M., Alvarez, J., Bilbao, J., & Olazar, M. (2018). Recent advances in the gasification of waste plastics. A critical overview. In *Renewable and Sustainable Energy Reviews* (Vol. 82, pp. 576–596). Elsevier Ltd. https://doi.org/10.1016/j.rser.2017.09.032.

Maqsood, M. & Altaf, E. A. (2023). Industrial ecology-design of closed loop system to minimize waste and reduce environmental impact. *International Journal of Innovative Research in Engineering and Management 10*(4), 114–120. https://doi.org/10.55524/ijirem.2023.10.4.14.

Merrington, A. (2017). Recycling of Plastics. In *Applied Plastics Engineering Handbook: Processing, Materials, and Applications: Second Edition* (pp. 167–189). Elsevier Inc. https://doi.org/10.1016/B978-0-323-39040-8.00009-2.

Mishra, S. & Upadhyay, R. K. (2021). Review on biomass gasification: Gasifiers, gasifying mediums, and operational parameters. *Materials Science for Energy Technologies*, *4*, 329–340. https://doi.org/10.1016/j.mset.2021.08.009.

Naderi K., E., Lotfian, S., Entezar Shabestari, M., Khayatzadeh, S., Zhao, C., & Yazdani Nezhad, H. (2023). A critical review of the current progress of plastic waste recycling technology in structural materials. In *Current Opinion in Green and Sustainable Chemistry* (Vol. 40). Elsevier B.V. https://doi.org/10.1016/j.cogsc.2023.100763.

Pandey, S., Karakoti, M., Dhali, S., Karki, N., SanthiBhushan, B., Tewari, C., Rana, S., Srivastava, A., Melkani, A. B., & Sahoo, N. G. (2019). Bulk synthesis of graphene nanosheets from plastic waste: An invincible method of solid waste management for better tomorrow. *Waste Management, 88,* 48–55. https://doi.org/10.1016/j.wasman.2019.03.023.

Pandey, U., Stormyr, J. A., Hassani, A., Jaiswal, R., Haugen, H. H., & Moldestad, B. M. E. (2020). Pyrolysis of plastic waste to environmentally friendly products. *WIT Transactions on Ecology and the Environment, 246,* 61–74. https://doi.org/10.2495/EPM200071.

Papari, S., Bamdad, H., & Berruti, F. (2021). Pyrolytic conversion of plastic waste to value-added products and fuels: A review. *Materials, 14*(10), 2586. https://doi.org/10.3390/ma14102586.

Papari, S. & Hawboldt, K. (2017). Development and Validation of a Process Model to Describe Pyrolysis of Forestry Residues in an Auger Reactor. *Energy and Fuels, 31*(10), 10833–10841. https://doi.org/10.1021/acs.energyfuels.7b01263.

Plastics Europe, Plastics – the Facts. (2022). https://plasticseurope.org/knowledge-hub/plastics-the-facts-2022/ (Accessed 2 October 2023).

Qureshi, M. S., Oasmaa, A., Pihkola, H., Deviatkin, I., Tenhunen, A., Mannila, J., Minkkinen, H., Pohjakallio, M., & Laine-Ylijoki, J. (2020). Pyrolysis of plastic waste: Opportunities and challenges. *Journal of Analytical and Applied Pyrolysis, 152,* 104804. https://doi.org/10.1016/j.jaap.2020.104804.

Ribeiro, M. C. S., Fiúza, A., Ferreira, A., Dinis, M. de L., Castro, A. C. M., Meixedo, J. P., & Alvim, M. R. (2016). Recycling approach towards sustainability advance of composite materials' industry. *Recycling, 1*(1), 178–193. https://doi.org/10.3390/recycling1010178.

Saebea, D., Ruengrit, P., Arpornwichanop, A., & Patcharavorachot, Y. (2020). Gasification of plastic waste for synthesis gas production. *Energy Reports, 6,* 202–207. https://doi.org/10.1016/j.egyr.2019.08.043.

Salaudeen, S. A., Arku, P., & Dutta, A. (2018). Gasification of plastic solid waste and competitive technologies. In *Plastics to Energy: Fuel, Chemicals, and Sustainability Implications* (pp. 269–293). Elsevier. https://doi.org/10.1016/B978-0-12-813140-4.00010-8.

Shah, H. H., Amin, M., Iqbal, A., Nadeem, I., Kalin, M., Soomar, A. M., & Galal, A. M. (2023). A review on gasification and pyrolysis of waste plastics. In *Frontiers in Chemistry* (Vol. 10). Frontiers Media S.A. https://doi.org/10.3389/fchem.2022.960894.

Singh, J. & Ordoñez, I. (2016). Resource recovery from post-consumer waste: Important lessons for the upcoming circular economy. *Journal of Cleaner Production, 134,* 342–353. https://doi.org/10.1016/j.jclepro.2015.12.020.

Singh, N., Hui, D., Singh, R., Ahuja, I. P. S., Feo, L., & Fraternali, F. (2017). Recycling of plastic solid waste: A state of art review and future applications. *Composites Part B: Engineering, 115,* 409–422. https://doi.org/10.1016/j.compositesb.2016.09.013.

Singh, P., Déparrois, N., Burra, K. G., Bhattacharya, S., & Gupta, A. K. (2019). Energy recovery from cross-linked polyethylene wastes using pyrolysis and

CO$_2$ assisted gasification. *Applied Energy*, *254*, 113722. https://doi.org/ 10.1016/j.apenergy.2019.113722.

Tejaswini, M. S. S. R., Pathak, P., Ramkrishna, S., & Ganesh, P. S. (2022). A comprehensive review on integrative approach for sustainable management of plastic waste and its associated externalities. In *Science of the Total Environment* (Vol. 825). Elsevier B.V. https://doi.org/10.1016/j.scitotenv. 2022.153973.

Thachnatharen, N., Shahabuddin, S., & Sridewi, N. (2021). The waste management of polyethylene terephthalate (PET) plastic waste: A review. *IOP Conference Series: Materials Science and Engineering*, *1127*(1), 012002. https://doi.org/10.1088/1757-899x/1127/1/012002.

Thompson, R. C., Moore, C. J., Saal, F. S. V., & Swan, S. H. (2009). Plastics, the environment and human health: Current consensus and future trends. *Philosophical Transactions of the Royal Society B: Biological Sciences*, *364*(1526), 2153–2166. https://doi.org/10.1098/rstb.2009.0053.

Vijayakumar, A. & Sebastian, J. (2018). Pyrolysis process to produce fuel from different types of plastic — A review. *IOP Conference Series: Materials Science and Engineering*, *396*(1), 012062. https://doi.org/10.1088/1757-899X/396/1/012062.

Zhao, X., Korey, M., Li, K., Copenhaver, K., Tekinalp, H., Celik, S., Kalaitzidou, K., Ruan, R., Ragauskas, A. J., & Ozcan, S. (2022). Plastic waste upcycling toward a circular economy. *Chemical Engineering Journal*, *428*, 131928. https://doi.org/10.1016/j.cej.2021.131928.

https://doi.org/10.1142/9789811297755_0005

Chapter 5

Review of Metal Recovery Processes in Liquid Extraction Systems

Vandana Kumari Jha[*,‡] and Soubhik Kumar Bhaumik[†]

*Department of Mechanical Engineering
Gyeongsang National University, Jinju-52828, South Korea
†Department of Chemical Engineering
Indian Institute of Technology (Indian School of Mines),
Dhanbad-826004, India
‡vandana97jsr@gmail.com

Abstract

This comprehensive review presents an in-depth examination of metal recovery processes within liquid extraction systems. Its advantages include high selectivity, efficient separation, and resource recovery potential from waste streams. Liquid extraction relies on the principle of partitioning metal ions between immiscible aqueous and organic phases, influenced by mass transfer kinetics and solvent choice. Beginning with an exploration of the foundational principles of liquid–liquid extraction, including phase equilibria and solute partitioning, the review offers insights into traditional methods while addressing associated challenges. Subsequently, the chapter discusses recent innovations, highlighting their advantages in terms of efficiency, selectivity, and reduced environmental impact. The integration of automation and miniaturization is discussed, showcasing their transformative effects on metal recovery precision and throughput. Additionally, the review highlights the critical importance of sustainability in modern metal recovery, emphasizing eco-friendly practices such as solvent recycling and the use of green extractants. Ongoing research and development efforts strive to enhance efficiency, selectivity, and environmental sustainability, ensuring that this technology will play a crucial role in meeting future critical metal demands. Overall, this chapter provides a comprehensive analysis of evolving metal recovery techniques within liquid extraction systems, offering valuable insights into their applications and potential to address the challenges of sustainable resource management.

Keywords: Liquid extraction systems, mass transfer, metal recovery, solute partitioning, liquid–liquid extraction.

1. Introduction

In recent years, the demand for rare and precious metals has surged due to their critical applications in various industries, ranging from electronics and renewable energy to medical devices. As traditional mining methods face limitations in meeting this demand while maintaining environmental sustainability, alternative approaches such as liquid extraction systems have gained prominence. Liquid extraction systems find applications in various industries, including hydrometallurgy (the extraction of metals from ores), chemical engineering, and pharmaceuticals. These systems offer advantages such as high selectivity for certain metals, efficient separation, and the potential to recover valuable resources from waste streams (Rydberg *et al.*, 2004; Perera & Stevens, 2011).

A liquid extraction system, often referred to as solvent extraction or liquid–liquid extraction (LLE), is a separation process that involves the transfer of one or more solutes from one liquid phase (*the feed or source phase*) into another liquid phase (*the solvent or extract phase*) (Hoogerstraete *et al.*, 2013). In recent years, there has been a growing interest in recovering valuable metals from electronic waste (e-waste). E-waste is a rapidly growing waste stream that contains a variety of valuable metals, including gold, silver, copper, and palladium. This method is commonly used for the separation and recovery of these metals from e-waste in an efficient and environmentally friendly manner from complex mixtures at a low cost (Mahandra *et al.*, 2017). In the context of metal recovery, a liquid extraction system is employed to selectively extract specific metals from solutions or slurries containing multiple elements. The process relies on the differential solubility of metals in different solvents or phases. The recovery of metals from such systems involves a complex interplay of chemical reactions, mass transfer, and fluid dynamics. By choosing an appropriate solvent and optimizing various parameters, such as temperature, pH, and flow rates, it becomes possible to selectively transfer the target metal ions from the feed solution into the solvent phase (Zhang *et al.*, 2018; Lerum *et al.*, 2020).

The field of liquid extraction systems has witnessed significant advancements over the years, with researchers exploring various aspects of metal recovery and separation. Numerous studies have investigated the fundamental principles governing solvent extraction, emphasizing the role of interfacial phenomena and mass transfer kinetics. Notably, pioneering works laid the foundation for understanding the thermodynamics of LLE, explaining how equilibrium distribution coefficients influence metal partitioning between phases (Blass *et al.*, 1986; Preston *et al.*, 1996; Perera & Stevens, 2011). More recent investigations have focused on the development of novel extractants and the optimization of extraction parameters to enhance selectivity and efficiency (Jada *et al.*, 2022; Jada *et al.*, 2023). Advancements in numerical simulation techniques have provided new avenues for studying complex extraction systems. Computational fluid dynamics (CFD) simulations have been applied to model the hydrodynamics and transport phenomena within extraction columns, shedding light on phase behavior and droplet dynamics (Khatir *et al.*, 2017; Vasilyev *et al.*, 2019). Additionally, kinetic models have been developed to describe the transient behavior of metal extraction processes, helping in the prediction of extraction efficiency over time.

However, while these studies have offered valuable insights, there remains a need for comprehensive integration of these studies. There are several important factors that are essential and need to be considered for future research, such as the critical importance of metal recovery, recent advancements in technology, and a wide range of applications. Therefore, a thorough and basic analysis of metal recovery processes in liquid extraction systems is required. This will help provide information and inspire future research and innovation, which are essential in the field of metal extraction and recovery.

This chapter is primarily focused on conducting an in-depth examination of the various methods and advanced technologies applied in the field of metal recovery through liquid extraction systems. The objective is to promote a more profound understanding of the intricate mechanisms underlying metal extraction processes. This undertaking involves the clarification of the fundamental principles governing different liquid extraction methods, including solvent extraction, ion exchange, and precipitation techniques. Furthermore, this chapter extends its scope by exploring current research trends and the emerging landscape of technologies within the realm of metal recovery. Through this exploration, the aim is to illuminate the present state of the field, foster better comprehension of its existing dynamics, and provide insights with the potential to stimulate innovative developments and future advancements.

2. Fundamental Principles of Liquid Extraction

LLE, a pivotal separation technique in metal recovery processes, operates on the principle of partitioning, where metal ions distribute themselves between two immiscible liquid phases. This typically involves an aqueous phase and an organic phase (generally hydrocarbons). Each phase exhibits affinity for specific metal ions. The equilibrium state achieved during the extraction process is governed by mass transfer kinetics, which shows a balance in the rate of transfer of metal ions between the two phases. The choice of organic solvent plays a crucial role, as different solvents exhibit varying affinities for specific metal species. Also, the control of pH is essential, as it influences metal ions in aqueous solutions and allows for selective extraction of target metals. Solvent extraction mechanisms, including ion exchange, solvation, and complexation, further govern the transfer process. By understanding these fundamental principles, liquid extraction systems can be designed and optimized for efficient and selective metal recovery.

3. Types of Liquid Extraction Systems

Liquid extraction systems are used to separate metal ions from aqueous solutions into an organic phase. The organic phase is then stripped of the metal ions, which are then recovered. The three main types of liquid extraction systems used for metal recovery are discussed in the following sections.

3.1 *Solvent extraction*

Solvent extraction is the most common type of liquid extraction system used for metal recovery. In solvent extraction, the metal ions are extracted into an organic solvent containing an extractant. The extractant is a chemical that selectively binds to the metal ions (Jha *et al.*, 2012). The principles behind solvent extraction are based on certain factors. The metal ions must form complexes with the extractant that are soluble in the organic phase. The organic solvent must be immiscible with water so that the two phases can be easily separated. The extractant must be selective for the target metal ions. The mechanisms of solvent extraction can be classified into two main types. The first type is ion exchange, where the metal ions are exchanged with other ions in the extractant. For example, an amine extractant can be used to extract copper ions from an aqueous solution by exchanging them with hydrogen ions. The second is chelation, where the metal ions form complexes with the extractant by sharing their electrons. For example, a hydroxyoxime extractant can be used to extract nickel ions from an aqueous solution by forming complexes with them.

3.2 *Ion exchange*

Ion exchange is a liquid extraction system that uses a solid ion exchanger to exchange ions from an aqueous solution. The ion exchanger is typically a resin or bead that contains functional groups capable of binding to ions. The metal ions are exchanged with other ions on the ion exchanger by forming bonds with the functional groups. The metal ions must have a higher affinity for the functional groups on the ion exchanger than the other ions present in the aqueous solution (Gotfryd *et al.*, 2006). The ion exchanger must be in contact with the aqueous solution for a sufficient amount of time for the ion exchange reaction to occur. The mechanisms of ion exchange can be classified into two main types: cation and anion exchange. In cation exchange, the metal ions are exchanged with other cations on the ion exchanger. For example, a cation exchange resin can be used to remove copper ions from an aqueous solution by exchanging them with sodium ions. In anion exchange, the metal ions are exchanged with other anions on the ion exchanger. For example, an anion exchange resin can be used to remove chromate ions from an aqueous solution by exchanging them with chloride ions.

3.3 *Precipitation*

Precipitation is a liquid extraction process that uses a chemical reagent to precipitate metal ions from an aqueous solution. The chemical reagent used is typically a base or a sulfide. The metal ions are precipitated by forming insoluble compounds with the chemical reagent. The metal ions must form insoluble compounds with the chemical reagent. The chemical reagent must be added to the aqueous solution in the correct amount to precipitate the metal ions without forming excess precipitate. In hydroxide precipitation, the metal ions are precipitated by forming insoluble hydroxides with a base. For example, copper ions can be precipitated by adding sodium hydroxide to the aqueous solution. In sulfide precipitation, the metal ions are precipitated by forming insoluble sulfides with a sulfide. For example, nickel ions can be precipitated by adding sodium sulfide to the aqueous solution. The selection of a liquid extraction system for metal recovery depends on a number of factors, including the type of metal being recovered, the composition of the aqueous solution, and the desired recovery rate. Solvent extraction is the most versatile and widely used liquid extraction system for metal recovery.

4. Metal Recovery Processes and Analysis

Following the leaching of valuable metals from metal waste (often using concentrated leachate obtained through acid dissolution), three primary post-processing techniques are currently used: solvent extraction, ion exchange, and precipitation. Solvent extraction is capable of handling large volumes and enabling the separation of rare earths based on their selective distribution (cations, complex anions, or neutral) between two immiscible liquid phases (Abreu *et al.*, 2014; Yoon *et al.*, 2016). Ion exchange excels at concentrating and purifying specific metal ions through selective binding to a resin. Precipitation offers the final isolation of the desired metal as a solid compound. Researchers like those employing the "dilute sulfuric acid solution-solvent extraction-precipitation technique" for extracting light and heavy rare earths from apatite ores demonstrate the effectiveness of these combined approaches (Battsengel *et al.*, 2018). Electronic products, packed with a diverse array of valuable metals, offer a tempting alternative to traditional ore mining. E-waste contains higher concentrations of metals such as Al, Cu, Fe, Pb, Ni, Zn, and precious metals (Au, Ag, Pt, and Pd) compared to virgin ores.

Additionally, it contains specialty metals such as Co, Se, and rare earth elements (REEs), which are essential for modern technologies (Marra *et al.*, 2019). Not only does e-waste recovery promise economic benefits, but it also reduces the environmental pressures associated with conventional mining. Bioleaching, however, remains a less explored technique for e-waste recycling compared to more established methods such as chemical and supercritical liquid leaching. Further research in bioleaching could unlock sustainable and environmentally friendly avenues for extracting valuable metals from this abundant and readily available resource. Therefore, effective metal recovery from different waste sources necessitates careful consideration of leaching methods and post-processing techniques to achieve optimal yield and sustainability.

5. Metal Recovery Processes and Analysis

Metal recovery processes in liquid extraction systems involve leaching, solid–liquid separation, liquid extraction, stripping, and precipitation (Treybal, 1963; Lo *et al.*, 1983). A detailed outline is shown in Figure 1 to help understand the visual representation of the LLE metal recovery process. The specific processes and techniques used can vary depending on the type of metal being recovered and the composition of the feedstock. Some of the most common processes and techniques include counter-current extraction, staged extraction, and column extraction. Liquid extraction systems offer a number of advantages over other metal recovery processes, including high selectivity, high efficiency, and versatility. However, they can also be expensive to install and operate, and they can generate hazardous waste. To optimize waste utilization and meet the growing demand for metals, researchers are actively exploring their extraction from different sources (Elderfield & Greaves, 1982; Zhao *et al.*, 2019). Table 1 illustrates the applications of metals recovered from LLE systems and their sources. Similar to traditional ore mining, recovering metals from industrial waste utilizes various techniques to extract elements such as gallium (used in LEDs and electronics), vanadium (for steel strengthening and batteries), and RREs (crucial for magnets in wind turbines and electric vehicles) (Bolch, 1980; Yang *et al.*, 2016). The average content of these metals in waste may be low, but the sheer volume — for example, the 21 million tons of RREs in global phosphogypsum — presents a significant opportunity (Wang *et al.*, 2015). Beyond resource recovery, extracting these metals also offers

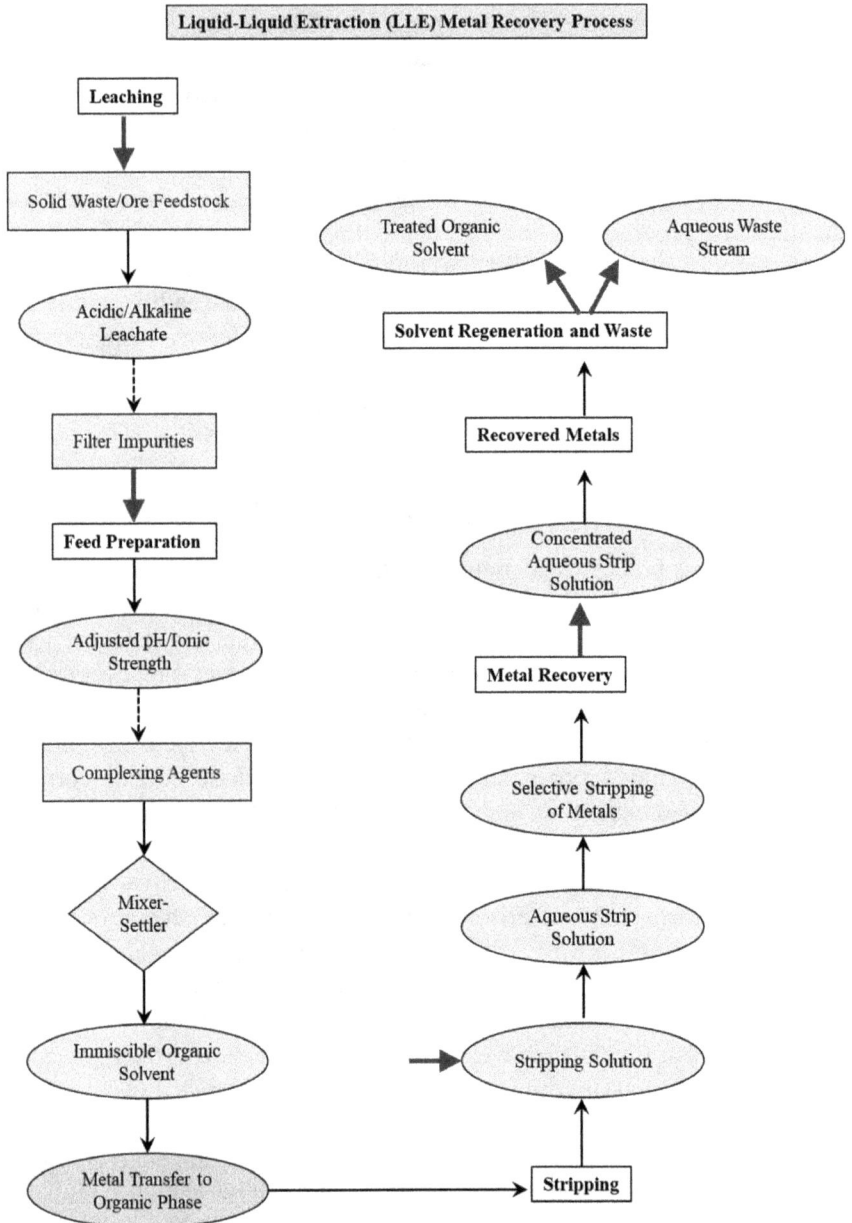

Figure 1.   Flowchart depicting the LLE metal recovery process.

Table 1. Applications of metals recovered from liquid–liquid extraction systems and their sources.

Metal	Source	Application examples
Gold (Au)	Printed circuit boards (PCBs), e-waste, ores	Jewelry, electronics, dentistry, medicine, aerospace
Silver (Ag)	PCBs, batteries, cables, e-waste, ores	Jewelry, electronics, photography, medicine, solar cells
Copper (Cu)	PCBs, wires, coils, e-waste, ores	Electrical wiring, plumbing, construction, machinery, transportation
Platinum (Pt)	PCBs, catalysts, autocatalysts	Jewelry, chemical catalysts, fuel cells, pacemakers, electronics
Palladium (Pd)	PCBs, catalysts, autocatalysts	Chemical catalysts, fuel cells, dentistry, electronics
Rare Earth Elements (REEs)	Magnets, batteries, e-waste, ores	Permanent magnets, electric motors, wind turbines, electronics, catalysts, phosphors
Lithium (Li)	Brines, pegmatites, e-waste	Rechargeable batteries, ceramics, glass, pharmaceuticals, nuclear fusion
Cobalt (Co)	Batteries, e-waste, ores	Lithium-ion batteries, superalloys, catalysts, magnets, pigments
Nickel (Ni)	Stainless steel, e-waste, ores	Batteries, stainless steel, electronics, catalysts, superalloys
Tantalum (Ta)	Capacitors, e-waste, ores	Tantalum capacitors, medical implants, aerospace, cutting tools

environmental benefits. Waste often contains hazardous elements such as arsenic and lead, which pose health and environmental risks (Yahorava *et al.*, 2016).

Recovering valuable metals not only reduces reliance on traditional mining but also helps to reduce these potential dangers through proper treatment and transformation. Furthermore, solid waste often comes in fine particle sizes, eliminating the need for pre-crushing, which is a significant energy and cost advantage when compared with standard ore processing. This makes the extraction process even more attractive. While challenges remain in optimizing efficiency and ensuring environmentally responsible practices, the potential for metal recovery from industrial waste is undeniable. It offers a sustainable solution for resource management, environmental protection, and meeting the ever-growing demand for critical metals in our modern world. Liquid extraction is widely applied in the mining and metallurgical industries to recover metals from ores and other feedstocks. Some examples of specific applications include the recovery of copper, zinc, gold, lithium, cobalt, and RREs (Khatir *et al.*, 2017). Overall, liquid extraction is a versatile and powerful technology for the recovery of metals from aqueous solutions (Zhang *et al.*, 2018). It is widely used in the mining and metallurgical industries, and it is also being used to recover metals from waste streams and other secondary resources.

6. Factors Affecting Metal Recovery

The pH of the aqueous solution is one of the most important factors that affects the efficiency of metal recovery processes. The pH affects both the solubility of the metal ions and the stability of the metal–extractant complexes. For example, copper ions are more soluble in acidic solutions, while they are less soluble in alkaline solutions. On the other hand, the metal–extractant complexes are more stable in alkaline solutions. This is because the hydroxide ions in alkaline solutions can deprotonate the extractant, making it more likely to form complexes with the metal ions. The temperature of the aqueous solution also affects the efficiency of metal recovery processes. The temperature affects the solubility of the metal ions and the stability of the metal–extractant complexes. In general, the solubility of metal ions increases with temperature. This is because the thermal energy at higher temperatures helps to break the bonds between the metal ions and the water molecules. The concentration of the metal ions in the aqueous solution also affects the efficiency of metal recovery processes. The concentration affects the driving force for the extraction process and the formation of the metal–extractant complexes. A good extraction agent should have high selectivity, high capacity, and low cost. However, it is difficult to find an extraction agent that meets all these criteria. Therefore, the selection of an extraction agent involves a trade-off between these factors.

7. Conclusions

In conclusion, this review has provided a comprehensive overview of metal recovery processes within liquid extraction systems. It is evident that these methods play a pivotal role in the extraction and purification of metals from various sources, with applications spanning from hydrometallurgy to environmental remediation. The diverse range of extractants and techniques discussed highlights the adaptability of liquid extraction systems to different metal recovery challenges. They find extensive application in mining and metallurgical industries, enabling the recovery of a wide spectrum of metals, including copper, zinc, gold, lithium, cobalt, and RREs. Three primary types of liquid extraction systems — solvent extraction, ion exchange, and precipitation — each offer their own set of advantages and disadvantages, with the choice depending on specific needs. Despite these considerations, liquid extraction systems provide significant benefits such as selectivity, efficiency, and versatility. It's crucial, however, to carefully consider factors such as pH, temperature, concentration, and extraction agents when selecting a system for a particular application. Ongoing research endeavors focus on developing novel liquid extraction systems that prioritize selectivity, efficiency, and environmental sustainability. In the years ahead, liquid extraction systems are anticipated to assume an increasingly vital role in metal recovery. With the ever-growing demand for metals, these systems will be indispensable in meeting this demand sustainably. Moreover, ongoing research and development efforts in this field continue to enhance the efficiency, selectivity, and environmental sustainability of these processes, promising a bright future for metal recovery through liquid extraction systems as they address the increasing demand for critical metals and sustainable resource management.

References

Abreu, R. D. & Morais, C. A. (2014). Study on separation of heavy rare earth elements by solvent extraction with organophosphorus acids and amine reagents. Minerals Engineering, *61*, 82–87.

Battsengel, A., Batnasan, A., Narankhuu, A., Haga, K., Watanabe, Y., & Shibayama, A. (2018). Recovery of light and heavy rare earth elements from apatite ore using sulphuric acid leaching, solvent extraction and precipitation, *Hydrometallurgy, 179*, 100–109. https://doi.org/10.1016/j.hydromet.2018.05.024.

Blass, E., Goldmann, G., Hirschmann, K., Mihailowitsch, P., & Pietzsch, W. (1986). Progress in liquid/liquid extraction. *German Chemical Engineering, 9*(4), 222–238.

Bolch Jr., W. E. (1980). *Solid Waste and Trace Element Impact*. University of Florida Press, Gainesville.

Elderfield, H. & Greaves, M. J. (1982). The rare earth elements in seawater. *Nature, 296*, 214–219.

Gotfryd, L. & Cox, M. (2006). The selective recovery of cadmium(II) from sulfate solutions by a counter-current extraction–stripping process using a mixture of diisopropylsalicylic acid and Cyanex® 471X. *Hydrometallurgy, 81*, 226–233.

Jada, N., Agrawal, N., Ganneboyina, S. R., & Bhaumik, S. K. (2023). Novel template digestive fabrication of multi-helical flow reactors. *Chemical Engineering Research and Design, 197*, 334–341.

Jada, N., Ganneboyina, S. R., & Bhaumik, S. K. (2022). Flow transitions in triple-helical microchannel involving novel parallel flow patterns. *Physics of Fluids, 34*(12), 124102.

Jha, M. K., Kumar, V., Jeong, J., & Lee, J. C. (2012). Review on solvent extraction of cadmium from various solutions. *Hydrometallurgy, 111–112*, 1–9.

Khatir, Z., Hanson, B. C., Fairweather, M., & Heggs, P. J. (2017). CFD analysis of liquid-liquid extraction pulsed sieve-plate extraction columns. *Computer Aided Chemical Engineering, 40*(2017), 19–24.

Lerum, H. V., Sand, S., Eriksen, D. O., Wibetoe, G., & Omtvedt, J. P. (2020). Comparison of single-phase and two-phase measurements in extraction, separation and back-extraction of Cd, Zn and Co from a multi-element matrix using Aliquat 336. *Journal of Radioanalytical and Nuclear Chemistry, 324*, 1203–1214.

Lo, T. C., Baird, M. H. I., & Hanson, C. (eds.) (1983). *Handbook of Solvent Extraction*. John Wiley & Sons, Florida, USA.

Mahandra, H., Singh, R., & Gupta, B. (2017). Liquid-liquid extraction studies on Zn(II) and Cd(II) using phosphonium ionic liquid (Cyphos IL 104) and recovery of zinc from zinc plating mud. *Separation and Purification Technology, 177*, 281–292.

Marra, A., Cesaro, A., & Belgiorno, V. (2019). Recovery opportunities of valuable and critical elements from WEEE treatment residues by hydrometallurgical processes. *Environmental Science and Pollution Research Intuition, 26*, 19897–19905. https://doi.org/10.1007/s11356-019-05406-5.

Perera, J. M. & Stevens, G. W. (2011). The role of additives in metal extraction in oil/water systems. *Solvent Extraction and Ion Exchange, 29*(3), 363–383.

Preston, J. S. & du Preez, A. C. (1996). Synergistic effects in solvent extraction systems based on alkyl salicylic acids 0.2. Extraction of nickel, cobalt, cadmium and zinc in the presence of some neutral N-, O- and S-donor compounds. *Solvent Extraction and Ion Exchange, 14*(2), 179–201.

Rydberg, J., Cox, M., Musikas, C., & Choppin, G. R. (2004). *Solvent Extraction Principles and Practice*, 2nd ed. Marcel Dekker, Basel.

Treybal, R. E. (1963). *Liquid Extraction.* McGraw-Hill, New York.

Vander Hoogerstraete, T., Onghena, B., & Binnemans, K. (2013). Homogeneous liquid–liquid extraction of rare earths with the betaine — Betainium bis(trifluoromethylsulfonyl)imide ionic liquid system. *International Journal of Molecular Sciences, 14*, 21353–21377.

Vasilyev, F., Virolainen, S., & Sainio T. (2019). Numerical simulation of counter-current liquid–liquid extraction for recovering Co, Ni and Li from lithium-ion battery leachates of varying composition. *Separation and Purification Technology, 210*, 530–540.

Wang, F., Zhang, Y., Liu, T., Huang, J., Zhao, J., Zhang, G., & Liu, J. (2015). A mechanism of calcium fluoride-enhanced vanadium leaching from stone coal. *International Journal of Mineral Processing, 145*, 87–93.

Yahorava, V., Bazhko, V., & Freeman, M. (2016). Viability of phosphogypsum as a secondary resource of rare earth elements. In *XXVIII International Mineral Processing Congress Proceedings.*

Yang, X., Zhang, Y., Bao, S., & Shen, C. (2016). Separation and recovery of vanadium from a sulfuric-acid leaching solution of stone coal by solvent extraction using trialkylamine. *Separation and Purification Technology, 164*, 49–55.

Yoon, H. S., Kim, C. J., Chung, K. W., Kim, S. D., Lee, J. Y., & Kumar, J. R. (2016). Solvent extraction, separation and recovery of dysprosium (Dy) and neodymium (Nd) from aqueous solutions: Waste recycling strategies for permanent magnet processing. *Hydrometallurgy, 165*, 27–43.

Zhang, L., Hessel, V., & Peng, J. (2018). Liquid-liquid extraction for the separation of Co(II) from Ni(II) with Cyanex 272 using a pilot scale Re-entrance flow microreactor. *Chemical Engineering Journal, 332*, 131–139.

Zhao, Y., Wang, S., Li, Y., Zhuo, Y., & Liu, J. (2019). Effects of straw layer and flue gas desulfurization gypsum treatments on soil salinity and sodicity in relation to sunflower yield. *Geoderma, 352*, 13–21.

Chapter 6

Sustainable Use of Nanofillers in the Textile Industry for Resource Recovery

Ermias Wubet Addiss* and Debleena Bhattacharya*,†

*Department of Chemical Engineering, Marwadi University,
Rajkot, Gujarat, India
†debleena.bhattacharya@gmail.com

Abstract

The prominent dyes present in the textile industry are azo, anthraquinone, and methylene blue (MB). These dyes are more resistant to degradation. The treatment imparted for the removal of color in the textile industry is divided into physicochemical and biological methods. Inexpensive agricultural wastes act as adsorbents for eliminating dyes from the solution. Adsorption methods are frequently employed to remove specific types of contaminants from water, particularly those that are difficult to biodegrade. For the removal of colors from wastewater, biological treatment and adsorption on activated carbon are currently gaining popularity. Though commercial activated carbon is the preferred adsorbent for color removal, its widespread use is constrained by its comparatively high cost, which prompted research on alternative non-conventional and affordable adsorbents. This review aims to compile the disparate information that is currently available on a variety of possibly affordable adsorbents for dye removal. It includes many recently published data that reveal that inexpensive biosorbents and nanofillers have excelled at removing dyes from textile wastewater.

Keywords: Nanofillers, textile wastewater, adsorption, natural waste.

1. Introduction

Huge quantities of colored wastewater frequently created in sectors such as rubber, paper, plastics, and textiles, which use dyes to give their products the right color, are dumped into natural streams with detrimental effects on the environment and human health (Ayalew *et al.*, 2020). Being one of the largest consumers of dyes, the textile sector accounts for 70% of its market, earns about US$1 trillion annually, represents 7% of all exports, and employs about 35 million people globally (Noman *et al.*, 2019). Most of the dye effluent produced worldwide, or over half, is emitted by the textile and apparel industries (Yaseen & Scholz, 2019). The dyeing process, which includes sourcing, size, desizing, bleaching, mercerization, tinting, printing, and finishing methods, is a crucial step in the production of textiles. The washing of colored or printed textiles results in large amounts of dye wastewater being released into the environment by the textile industry (Al-Gheethi *et al.*, 2022). It is reported that over 700,000 tons of dyestuff are produced annually, with over 100,000 commercially accessible dyes (McMullan *et al.*, 2001). Methylene blue (MB) is a common dye mostly used by industries involved in textiles, paper, rubber, plastics, leather, cosmetics, pharmaceuticals, and foods. (Mohammed *et al.*, 2014). It is acknowledged that the color of water has a significant impact on how the public perceives its quality. The color is the first component in wastewater that can be identified. Less than 1 ppm for some dyes indicates the presence of colors in water, which is extremely apparent and undesirable (Robinson *et al.*, 2001).

Recently, the lack of access to clean water in society has been linked to the pollution of water bodies by untreated MB dye effluents produced by industry (Sleiman *et al.*, 2007). This happens frequently in underdeveloped nations, where large quantities of wastewater are dumped into the environment without proper control (Zaghbani *et al.*, 2008). Since MB dye has become a threat to the physical environment, scientists have been looking for solutions to clean up the environment and get rid of this dye. To remove this dye from the environment, a variety of treatment techniques have been used, including biological techniques (using enzymes and microorganisms), chemical techniques (using sophisticated oxidation processes), and physicochemical techniques (mainly adsorption) (De Oliveira *et al.*, 2011). The adsorption techniques have been studied and used in relation to the removal of dyes from the environment and include, but are not limited to, the use of bamboo stalk charcoal, sugarcane bagasse, potato peel, and biochar (Ahmad *et al.*, 2020), activated mango leaf (Tiwari *et al.*, 2015), onion skin (Saka *et al.*, 2011), walnut shell (Quansah *et al.*, 2020), activated carbon (Gohr *et al.*, 2022), and sawdust.

The removal of dye from textile effluents can be partially accomplished using conventional wastewater treatment techniques, such as sedimentation, chemical flocculation and coagulation, filtration, and aeration. However, these methods have a number of drawbacks, including the production of harmful byproducts, high energy costs, unpleasant odors, and the need for a sizable treatment area. To close these gaps and enhance the quality of textile wastewater treatment prior to its final release into the environment, researchers must identify more effective solutions. In recent years, there has been an increased interest in creating industrial textile wastewater treatment procedures using more effective methods (Al-Gheethi *et al.*, 2022).

A number of physical, chemical, and biological decolorization techniques for textile waste water treatment have been reported during the past three decades, but only a small number have been approved by the paper and textile sectors (Ghoreishi *et al.*, 2003). Adsorption, which can be used to remove various kinds of coloring materials, is the method of choice and produces the greatest results among the many dye removal processes (Ho *et al.*, 2003). Due to its great efficiency compared to biological methods that have several limitations, the adsorption method was frequently used to remove dye from wastewater. For some types of dyes, using the biomass of live organisms for dye adsorption or breakdown is ineffective (Noman *et al.*, 2021). Due to the high effectiveness of effortlessly removing colors and the absence of toxic by-products, the adsorption method employing agricultural wastes is considered more effective. Products made from agricultural waste have little to no commercial value, making their disposal a challenge. For the removal of numerous contaminants from wastewater, the use of agricultural wastes as adsorbents is crucial. The agricultural waste has a porous, loose structure as well as functional groups such as hydroxyl and carboxyl that efficiently aid in the dyes' adsorption (Dai *et al.*, 2018).

The current chapter highlights the various environmentally friendly methods for the removal of MB to improve the efficiency of textile wastewater treatment. The use of agricultural wastes as adsorbents for removing colors has not yet been thoroughly examined. The agricultural wastes that were employed as adsorbents to extract MB in earlier studies are examined and analyzed in this chapter. This review focused on the removal of MB from aqueous solutions by various agricultural wastes and also the use of nanofillers as adsorbents. It presents a wide range of adsorption results to summarize current research on the use of agricultural waste materials as dye adsorbents in wastewater.

2. Treatment Methods for Dye Wastewater

Since the 1990s, equalization and sedimentation processes have been used in dye removal techniques as the first water purification steps (Geer *et al.*, 2018). To provide more efficient removal approaches, activated sludge and dye-degrading filter bed operations were improved. The most popular color removal techniques used today are adsorption, biological therapy, electrochemical treatment, and the advanced oxidation process (Figure 1).

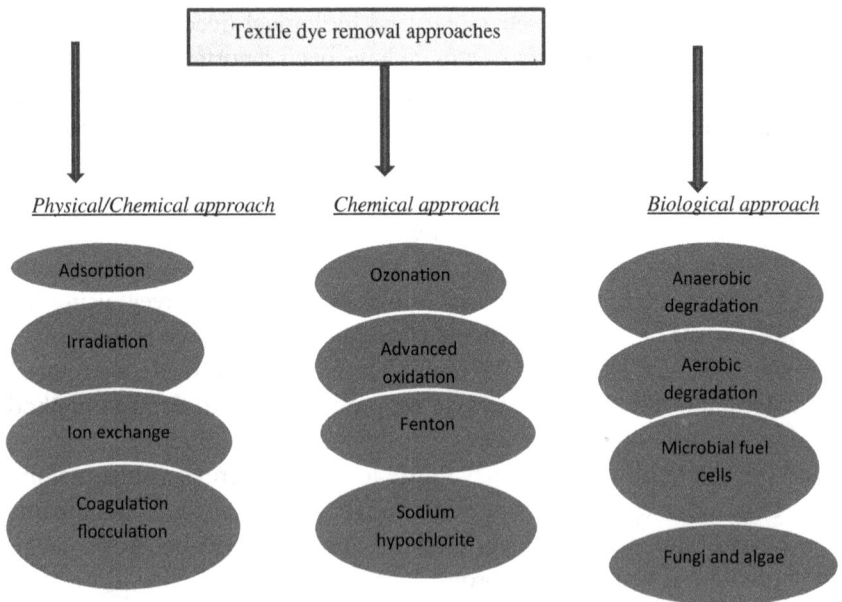

Figure 1.   Approaches for textile wastewater color removal.

Physical, chemical, and biological treatments are the hallmarks of the dye wastewater treatment systems currently in use. Due to the complex existence of effluents from numerous procedure sectors, there is no standalone approach that has proven appropriate to satisfy the regulatory criteria. More frequently, a combination of many different treatment methods is employed to provide treated water of the best possible quality while remaining cost-effective (Crini & Lichtfouse, 2019). Different textile wastewater treatment methodologies, with their merits and demerits, are shown in Table 1. Adsorption is superior to other methods for treating dye wastewater, offering important advantages such as low cost, accessibility, profitability, simplicity of use, and dependability (Al-Gheethi *et al.*, 2022).

As a way to remove color, biological techniques such as anaerobic-aerobic digestion (Manavi *et al.*, 2017), enzyme degradation, and microbial

Table 1. Merits and demerits of various treatment approaches.

Textile effluent treatment method	Merits	Drawbacks	Reference
Flocculation/ Coagulation	• Major COD and BOD reduction • A broad variety of chemicals are offered for sale on the market • COD and BOD levels significantly reduced.	• Includes the addition of non-reusable chemical substances (coagulants, flocculants, aid chemicals) • An increase in the volume of sludge produced • A large quantity of leftover aluminum	Crini and Lichtofouse, 2019
Adsorption and biosorption	• This method needed oversight and money for upkeep • Enabled carbon cost • Chemical synthesis to increase their adsorption capability • Regeneration issues	• This technique needed monitoring and money for upkeep. • Enabled carbon cost • Chemical derivation to boost their capacity for adsorption • Regeneration issues	Birniwa *et al.*, 2022

(*Continued*)

Table 1.   (*Continued*)

Textile effluent treatment method	Merits	Drawbacks	Reference
Membrane filtration	• Removes all dye types • It is can be easily coupled with other processes and operations with no additives and chemicals. • Effective for water recovery and reuse; little space requirement	• Equipment costs for producing concentrated sludge might be considerable • High energy demands • Different membrane filtering system designs exist • High maintenance, • Rapid membrane obstruction	Sivarajasekar and Baskar, 2015
Biodegradation	• Reusable and only suitable for azo dye removal • Reusable, highly effective, and safe • A range of colors can be removed concurrently • It is considered rapid when the dye wastewater is decolored in as little as 30 hours	• Produces high quantities of sludge. • Unreliable amount of enzyme production occurs • requires a nitrogen confined area to grow – Functional for only a limited number of dyes	Praveen *et al.*, 2021
Ion exchange	• Produce high quality of water • Regeneration: No loss of adsorption	• Limited number of dyes that it is effective for • Power sensitive to effluent pH	Sivarajasekar and Baskar (2015)

biomass adsorption are frequently used (Bayomie *et al.*, 2020). The typical practice is to implement a combination of aerobic and anaerobic procedures prior to discharge into the natural water system. The approach was chosen as the standard for dye removal. The other dye removal methods include adsorption by microorganism biomass and enzyme degradation.

Srinivasan *et al.* (2019) investigated the relationship between an enzyme and a textile dye. Remazol Blue RGB, Remazol Red RGB, Joyfix Red RB, Reactive Red F3B, Turquoise CL-5, and Joyfix Yellow MR are six of the textile dyes that were utilized. The different affinities, active sites, and enzyme/dye interactions were elucidated via the docking analysis. The oxido-reductive enzymes azoreductase and laccase were found to be responsible for the degradation of phthalocyanine and anthraquinone, according to the docking analysis. The authors also came to the conclusion that the degradation process involves the systematic destruction of azo bonds and a ping-pong mechanism involving free radicals.

The enzymatic processes are reliable and adequate, nontoxic, efficient, cheap, and reusable. However, since the biological method deals with living organisms, the major disadvantage is the growth rate of the microorganism and the consequences for the discharge of the treated textile wastewater with these organisms into the environment. Tan *et al.* (2019) used the halotolerant yeast Candida tropicalis SYF-1 to study the breakdown of the azo dye. The findings showed that SYF-1 yeast was able to decolorize all six azo dyes, with Acid Red B (ARB) dye being the most successfully eliminated. The optimal conditions for ARB dye decolorization included 0.6 g/L of ammonium sulfate, 4.0 g/L of glucose, 30 g/L of NaCl, a pH of 7.0–8.0, a temperature of 30°C, and a rotating speed of about 160 rpm. Bankole *et al.* (2018) employed *Achaemtomium strumarium* to speed up the process of acid red 88 dye removal. The fungal strain was acclimatized to 10–80 mg. When the adsorption capacity was 36.75 mg g1, the removal effectiveness was 99%. In order to create an ultrafiltration membrane, Isik *et al.* (2019) treated actual textile wastewater using the filamentous fungal *Aspergillus carbonarius* M333. The scientists considered growth periods of 3, 5, and 9 days to measure the growth of fungi. The ultrafiltration membrane made from fungi resulted in a 91% decolorization and a 73% reduction in COD, according to the research. Through the mass transfer technique, the physical dye removal method becomes an easy operation. Reverse osmosis, ion exchange, adsorption, radiation, nanofiltration, and membrane filtration are the processes frequently used to remove the dye (Holkar *et al.*, 2016).

The physical process uses fewer chemicals than biological and chemical procedures combined. The method's percentage of dye removal ranges from 86.8 to 99%. The adsorption procedure is repeated repeatedly until the adsorbent is utilized in a manner similar to the enzyme degradation procedure (Zietzschmann *et al.*, 2016). Due to this method's efficacy, some adsorbents may be expensive, which is a disadvantage of the adsorption approach. Utilizing inexpensive adsorbents, such as agricultural waste for adsorption, can overcome this problem (Hethnawi *et al.*, 2017).

In addition, there are microwave treatment (Remya *et al.*, 2011), liquid-liquid extraction (Sulaiman *et al.*, 2019), vacuum membrane distillation (Dragoi *et al.*, 2021), phytoremediation (Mustafa *et al.*, 2021), ultrafiltration (Yu *et al.*, 2020), and nanofiltration. However, studies have revealed that the majority of these traditional technologies are characterized by several downsides, such as being pricey, expensive electricity use, and high amounts of toxic waste formation. In this compiled list of MB removal techniques, these drawbacks and other comparisons will be examined and outlined (Oladoye *et al.*, 2022).

The disadvantage of the adsorption approach is that some adsorbents may be expensive. This can be resolved by using inexpensive adsorbents, such as agricultural waste, for adsorption (Hethnawi *et al.*, 2017).

Ishak *et al.* (2022) converted tamarind seed from a waste product with a negative value to activated carbon. In order to carry out the adsorption of synthetic colors, $ZnCl_2$ was used to chemically activate the specific agricultural waste to increase its surface area and improve its porosity. To physically activate the tamarind seeds' carbon, the carbonization process involved burning the seeds for one hour at 300°C, letting them cool for 24 hours, and then washing them with HCl. The effects of variables such as contact time, initial concentration, absorbance dosage, and pH on the adsorption of the dyes by tamarind seed activated carbon were investigated. According to the experimental results, both synthetic dyes showed a Langmuir isotherm during adsorption, with an R^2 correlation value of 0.9227 for MB and 0.6117 for other synthetic dyes (Reactive black 5). In the meantime, it was discovered that a pseudo-second-order model accurately predicted the rate of adsorption of MB and Reactive black 5 (RB5) by tamarind seed activated carbon. Hameed *et al.* (2016) used oil palm ash as an effective adsorbent for MB. The oil palm was activated using mesoporous zeolite. The prepared zeolite–activated carbon (Z–AC) composite was characterized by X-ray diffraction, Fourier transform infrared spectroscopy, BET surface area and pore structural analysis, and scanning electron microscopy. The adsorption performance of Z–AC for MB removal was studied utilizing a batch technique. On the adsorption of MB on Z–AC, the effects of starting dye concentration (25–400 mg/L), temperature (30–50°C), and pH (3.0–13.0) were investigated. It was discovered that pseudo-second-order kinetics accurately predicted the adsorption process. At 30, 40, and 50°C, respectively, the Z–AC composite's highest adsorption capacities for MB were 143.47, 199.6, and 285.71 mg/g. These findings demonstrate that the Z–AC composite may serve as a foundation for other low-cost composites that can be employed as dye adsorbents.

The chemical dye removal techniques showed a high efficiency to remove dye between 88.8 and 99% instead of physical and biological approaches (Oladoye *et al.*, 2022). Solid chemical reactions with the adsorbed substance lead to this kind of adsorption. As opposed to physical absorption, adhesion forces are typically stronger. For a material to be physically adsorbed at high temperatures, chemical absorption may be detected. The heat emitted during the chemical adsorption process is often high and about a chemical reaction. On catalysts, chemical adsorption is significant. Coagulants and flocculants are the main chemical agents used to remediate wastewater from dyeing. It entails adding materials to the effluent, such as calcium, aluminum, or ferric ions, in order to generate flocculation. Wang *et al.* (2018) have also reported the utilization of other substances for chemical reactions, including ferric sulfate and a few synthesized organic polymers.

Midway through the 1990s, the electrocoagulation technique was created. It has a number of significant benefits as a technique that removes color effectively. To dissolve the metal sheets in the wastewater, a direct current source is used in the effluent between metal electrodes, such as iron and aluminum (Zazou *et al.*, 2019).

Generally speaking, chemical treatment is economically viable and effective, but a key disadvantage is that chemical costs are high and prices fluctuate on the market due to strong demand and the rate at which chemicals are manufactured. Additionally, even if chemical treatment is effective, one major drawback is the generation of pH-dependent sludge at the end of the process, which causes disposal issues.

3. Methylene Blue Dye

For thousands of years, people have used color to communicate. According to scientists, the first recorded use of dyes and dye products occurred 180,000 years ago. About 4000 years ago, organic dyes were used for the first time. The discovery of blue in the Egyptian pyramids served as the foundation for this information. Chromo forms, which are responsible for producing color, and exochromes, which not only work in conjunction with Chromo forms to produce color but also dissolve dye in water and absorb color molecules into fiber surfaces, make up the two main components of color molecules (Rahimian *et al.*, 2020).

Dyes are colored organic compounds used in a variety of sectors, including textiles, leather, paper, plastic, pharmaceuticals, cosmetics, and foods. They can also be used on hair, fur, oil refinery products, grease, and other organic materials like hair and fur. Acidic, basic, reactive, specific, azo, vat, caustic, dispersion, and sulfur dyes are frequently used in the garment manufacturing industries (da Silva *et al.*, 2020). The main category of dyes utilized in industrial applications is azo dye derivatives (Abbas *et al.*, 2020). The azo dye group is the largest, followed by the anthraquinone dye group. These colors are categorized as natural colors found in lichens, plants, and microorganisms. Indigo dye is a substance with a distinctive blue hue that is removed from plants and used in cosmetics, semiconductors, and medications (Al-Gheethi *et al.*, 2022). The dyes can generally be divided into three groups: cationic, anionic, and non-ionic dyes. The direct, acidic, and reactive dyes are anionic dyes.

MB is a common cationic dye that is used to coat paper as well as dye cotton, silk, leather, and wool (Dahlan *et al.*, 2019). Cationic dyes are known to be more harmful than other types of dyes when compared to dyes that are classified differently. MB is no more hazardous than any other cationic dye. However, prolonged exposure to MB can cause serious health problems such as vomiting, accelerated heartbeat, cyanosis, shock, limb paralysis, jaundice, eye burns, tissue necrosis, and mental confusion (Feng *et al.*, 2012). Wastewater that has been contaminated with MB needs to be removed before it is released into a water environment in order to lessen its detrimental effects on people.

The IUPAC name of [7(dimethylamino) phenothiazin-3-ylidene]-dimethyl azenium; chloride It has a molecular formula of $C_{16} H_{18} ClN_3 S$ and a molecular weight of 319.851 g/mol. The other physicochemical parameters of MB dye are reported in great depth in Table 2 (Al-Gheethi *et al.*, 2020). The greatest light wavelength that MB can absorb is close to 670 nm. The specifics of absorption are influenced by a variety of variables, such as protonation, adsorption to other substances, and metachromasia, which results in the production of dimers and higher-order aggregates in response to concentration and other interactions (Cenens *et al.*, 1988).

MB dye Figure 2 is a common blue, cationic, and thiazine dye that has been extensively used in the textile industry as a fiber coloring agent (Ahmad *et al.*, 2020), as well as in the field of medicine as staining agents, as well as for prophylactic and therapeutic applications (Ponraj *et al.*, 2017).

Table 2.   Extensive details of the physicochemical properties of MB dye.

Sr. No	Parameters	Name/Values
1.	Aqueous pH	2.0–3.5
2.	Degree of solubility	3.55%
3.	Maximum wavelength of absorption (λ_{max})	664 nm
4.	Another name	Swith blue
5.	Ionization	Basic
6.	Color index name	Basic blue
7.	Color index number	52015
8.	Molecular weight	319.85 g/mol

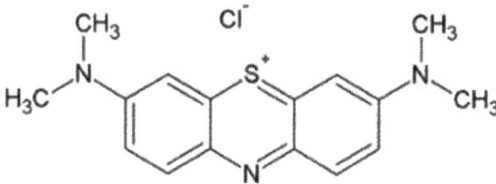

Figure 2.   Chemical structure of MB dye.

The uses for MB are countless. Medically, MB, which was tried around the end of the nineteenth century, was one of the first antimalarial medications to be created. Falciparum malaria was treated with it in addition to amodiaquine, and it was reported to be successful, particularly in children and adults from Africa (Bosoy *et al.*, 2008). The treatment of children and adults with methemoglobinemia, a blood condition in which an abnormal level of methemoglobin is formed, is advised for the use of MB in medicine (Grayling *et al.*, 2003). Other clinical uses for MB include treating hypotension associated with various clinical conditions, using it to treat hypoxia, and treating hyperdynamic circulation in liver cirrhosis, among others.

In the industrial setting, MB is predominantly used in the garment and textile sectors to dye a variety of textiles (Sharma *et al.*, 2022). Papers and leather can also be dyed with it. Surprisingly, MB has been employed as an indirect food additive in the food sector. Additionally, MB dye is utilized in medicine, microbiology, and diagnostics as a sensitizer in the photo-oxidation of organic compounds (Gupta *et al.*, 2016), as well as in aquaculture to cure a variety of fish diseases (Bharti *et al.*, 2019). As a result, it is undeniably a pertinent dye based on the uses of MB that have been discussed. However, MB's significant toxicity and refractory nature (Dahlan *et al.*, 2019) have been shown to be damaging to human health at a certain concentration, making it a potential concern to the ecosystem and to human health (Pavan *et al.*, 2008).

3.1 *Removal of MB textile dye by adsorption*

MB has a high solubility in water at ambient temperature (Russo *et al.*, 2016) and is famously challenging to biodegrade and remove from wastewater using basic conventional treatment methods (Oyarce *et al.*, 2020). Environmentally speaking, it is crucial to remove MB from effluent wastes in order to stop the detrimental impact it has on both human health and the environment (Zhu *et al.*, 2009). By regulating the dosage to be given, it is possible to regulate the use of MB in medicine. Therefore, wastewater and other industrial effluents are the main targets of the removal strategies.

The adsorption technology has been widely and effectively used to remove MB from wastewater (Babu *et al.*, 2019). This technology mainly depends on agricultural waters and nanofillers; Nanofillers are tiny particles that have a high surface area to volume ratio, making them ideal for adsorbing pollutants from wastewater. They can be made from a variety of materials, including metals, oxides, and carbon-based materials. One of the most commonly used nanofillers in textile wastewater treatment is graphene oxide, which has been shown to effectively remove dyes and other pollutants from water. Other types of nanofillers that are used in wastewater treatment include titanium dioxide, iron oxide, and silver nanoparticles.

The process results in highly treated water quality (Hethnawi *et al.*, 2017). Adsorption is the separation of a material from one phase and the accumulation or concentration of that substance at the surface of another. The substance condensed or adsorbed at the surface of the adsorbing phase is known as the adsorbate. Absorption, a process in which material is moved from one phase to another (for example, liquid), interpenetrates the second phase to form a "solution," and is thus distinct from adsorption. Both of these processes are generally referred to as "sorption."

Van der Waals forces and electrostatic interactions between adsorbate molecules and the atoms that make up the adsorbent surface are the fundamental causes of physical adsorption. As a result, surface characteristics like surface area and polarity are used to first describe adsorbents. The use of expensive adsorbents places a cap on the amount of adsorption that can occur. Adsorption would become one of the most cost-effective strategies available for eliminating dye effluents if low-cost adsorbents with high adsorption effectiveness were to be discovered (Babu *et al.*, 2019). Adsorption is a remarkable dye removal technique because it doesn't necessitate any particular removal conditions or pre-treatment of the effluent. Adsorbents typically consist of high specific surface area (200–2000 m^2/g) microporous materials.

3.2 *Agricultural wastes and nanofiller adsorbents*

Methylene blue is a common dye used in textile industries which poses a serious environmental threat due to its persistence in water bodies. Several studies have explored the potential of nanofiller agricultural wastes to remove MB from wastewater due to their high surface area and high adsorption capacity. Raw agricultural solid wastes like leaves, fibers, fruit peels, seeds, and other items have been employed as adsorbents, as well as waste products from the forestry sector like sawdust and bark. These materials are readily available, have favorable physicochemical properties, and could serve as adsorbents (Rafatullah *et al.*, 2010). Nanofillers and activated carbon are two types of materials that are commonly used in textile wastewater treatment due to their unique properties and effectiveness in removing pollutants from water. Agricultural wastes are often used as source materials to create activated carbon and biochar since they are inexpensive commodities with a high carbon concentration (Neme *et al.*, 2022). Despite the fact that agricultural wastes have no commercial value, turning them into active carbon helps to provide new, inexpensive activated carbon that is highly effective at removing colors from contaminated streams.

The capacity of a wide variety of affordable, low-cost adsorbents to remove various types of pollutants from wastewater has been carefully examined (Hashem, 2020). The adsorption capacities of different agricultural wastes and nanofillers are summarized in Table 3.

Researchers are working to generate more generally available varieties of adsorbents that are accepted, trustworthy, affordable, and easily accessible.

Table 3.   Different biosorbents and their respective adsorptive capabilities.

Adsorbent dye	Adsorbate dye	Adsorption capacity	Percentage removal (%)	Reference
Malachite green	Neem sawdust	4.354 mg/g	—	Odoemelam *et al.*, 2018
Methylene blue	Modified zeolitic imidazolate framework (ZIF-8) nanofiller	534 mg/g.	99.8%	Yang *et al.*, 2020
Congo red	Dried roots of Eichhornia crassipes	—	96%	Dbik *et al.*, 2020
Basic yellow 21	Wheat straw	71.43 mg/g	—	Hassanein *et al.*, 2009
Crystal violent	Grapefruit peel powder.	—	96% removal	Saeed *et al.*, 2010
Methylene blue	Graphene oxide nanofillers	96.2%	313 mg/g	Qu *et al.*, 2019
Reactive Red 23 dye	Cistus ladaniferus shell	97.046%	265.96 mg/g	El Farissi, *et al.*, 2021
Methylene blue	OIL-PALM FIBER	862.07 mg/g	—	Wang *et al.*, 2021
MB	Wheat shells	46.6 mg/g	95%	—
Reactive red 141	Sesame waste	27.55 mg g−1	—	Sohrabi *et al.*, 2016
Direct red dye effluent	Rice Husk	13 mg/g	—	Abdelwahab *et al.*, 2005
Malachite green indigo carmine	Cinnamon bark	—	84%	Güler *et al.*, 2021
Methylene blue	Corn cob	q_{max} = 333 mg/g.	—	Choi *et al.*, 2019

Low-cost adsorbents are those that need little processing, are readily available, or are leftovers or by-products from industry. Agricultural wastes are utilized as adsorbents to remove the dyes since they are inexpensive or free to use. Due to the growing food industry and global population, these wastes are produced in significant amounts.

One of the greenest and most often used methods for removing dye-stuff from industrial effluents is bio-wastes in dye adsorption, which is also regarded as waste management (Pavan *et al.*, 2008).

Jawad *et al.* (2018) investigated the efficacy of agricultural sorbent by activating banana peel with H_2SO_4. Acid-treated banana peel (ATBP) was created as a possible adsorbent for MB in an aqueous solution. According to the research, acid-treated banana peel (ATBP) activated with H_2SO_4 serves as a cheap adsorbent for the removal of MB dye from aqueous solutions. The findings of the adsorption trials show a maximum adsorption capacity (q_{max}) of 250 mg/g, better describing the adsorption results at equilibrium. According to the thermodynamic characteristics, the adsorption process is endothermic in nature and spontaneous. According to the findings, ATBP is a powerful adsorbent for MB adsorption. It is safe, responsible for the environment, and cost-effective to remove dye using agricultural waste as an adsorbent (Wu *et al.*, 2021). Agricultural waste is a viable resource for environmental technologies since it is renewable and can be utilized to clean water and wastewaters (Lewoyehu *et al.*, 2021).

Shakoor *et al.* (2016) investigated the possibility of citrus limetta peel (CLP) as a cheap agricultural waste adsorbent for the eradication of MB dye. Batch adsorption experiments were carried out to determine the effects of contact time, initial dye concentration, adsorbent dosage, pH, and temperature on adsorption. It was discovered that it had a maximum adsorption capacity of 227.3 mg/g. According to the results, CLP is a very efficient and reasonably priced adsorbent for the removal of colors from wastewater.

Gezer *et al.* (2018) studied the possible use of carob powder as a substitute adsorbent for the removal of MB from wastewater. Investigations were made into the experiment's pH, ultrasonic frequency, particle size, contact time, temperature, and initial MB dye concentration. For the dye MB, the equilibrium time was 267.63 min. The outcomes suggested that the adsorption process appeared to be significantly influenced by physisorption. The rate of adsorption of MB on carob powder was shown to be exothermic and to decrease with increasing temperature. The findings of the regression analysis showed that the experimental data fit well into the non-linear model, with correlation coefficients (R^2) of 0.8899 and 0.9830, respectively. The value of maximum adsorption was found to be 256,435.5 mg/g. When compared to other results that have been published in the literature, this finding seems to be significant. The study's findings revealed that carob beans can be employed as an alternate adsorbent. The yellow passion fruit (*Passiflora edulis* Sims. f. flavicarpa Degener) peel was examined (Pavan *et al.*, 2008). MB was removed from aqueous solutions using a prospective low-cost alternative adsorbent. Batch adsorption isotherms at room temperature were used to study the adsorption of MB onto this natural adsorbent. A study was done on the impact of pH and shaking time on adsorption capacity. The adsorption of MB was more successful at an alkaline pH.56 hours of contact time at 25°C were necessary to achieve maximal adsorption. An alternate adsorbent to extract MB from aqueous solutions is the yellow passion fruit peel.

One study by Yang *et al.* (2020) investigated the effectiveness of a modified zeolitic imidazolate framework (ZIF-8) nanofiller in removing MB from wastewater. The results showed that the modified ZIF-8 was able to remove up to 99.8% of MB from water, with an adsorption capacity of 534 mg/g. The study also found that the adsorption capacity of the modified ZIF-8 was significantly higher than that of the unmodified ZIF-8, demonstrating the importance of surface modification in enhancing the performance of nanofillers for wastewater treatment.

Dahlan *et al.* (2018) used sago waste (SW) produced for the removal of MB. The waste showed a high adsorption capacity for MB, with a Langmuir maximum adsorption capacity of 158 mg/g. After three cycles with a total removal of 171 mg/g of MB, the reusability test with 50 mg/L of MB revealed that the SW could be used again. The equilibrium adsorption data of SW were found to fit the Langmuir isotherm model quite well, proving the monolayer limitation of the adsorption process. On the other hand, it was discovered that the adsorption kinetics of MB onto SW were effectively fitted into a pseudo-second-order model, indicating that the adsorption mechanism was chemisorption. According to the intraparticle diffusion model's findings, SW displayed two-stage intraparticle diffusion.

Another study by Qu *et al.* (2019) investigated the use of activated carbon and graphene oxide nanofillers for removing MB from wastewater. The study found that graphene oxide had a higher adsorption capacity than activated carbon, with a maximum adsorption capacity of 536 mg/g compared to 313 mg/g for activated carbon. The study also found that the adsorption capacity of graphene oxide was influenced by factors such as pH, contact time, and initial dye concentration.

Yanfang *et al.* (2012) worked on the development of a promising and competitive bioabsorbent with an abundance of sources, a low price, and environmentally friendly characteristics to remove cationic dye from wastewater. In various operational circumstances, swede rape straw (*Brassica napus* L.) modified by tartaric acid (SRSTA) was produced, described, and used to remove MB from an aqueous solution (including MB initial concentrations, adsorbent dose, etc.). SRSTA's highest MB adsorption capacity of 246.4 mg/g was equivalent to the outcomes of several previously investigated activated carbons. The more functional groups, such as carboxyl groups, that are present on the surface of SRSTA, the higher the dye's capacity for adsorption. Discussions also included the adsorption mechanism. The findings suggest that SRSTA is a valuable and promising absorbent for removing MB from wastewater. Subramaniam *et al.* (2015) revealed that for the purpose of removing the dye MB from an aqueous solution, activated carbon made from cashew nut shells (CNS), an agricultural waste, was an effective adsorbent at optimum values of MB dye. The optimal values of pH, adsorbent dose, beginning dye concentration, and time were discovered to be 10, 2.1846 g/L, 50 mg/L, and 63 min for complete removal of MB dye, respectively.

Kuntari *et al.* (2018) utilized bamboo leaf wastes as an adsorbent material for MB adsorption. The characteristics of adsorption were studied in the batch adsorption procedures. High correlation coefficients were used to characterize the kinetics of MB adsorption using the pseudo-second-order reaction rate model. The experimental results were well-fit by the Freundlich model. Ash from bamboo leaves was shown to have a high MB adsorption. At a contact time of 20 min, a starting MB concentration of 10 mg/L, an initial pH of 6, and an adsorbent weight of 0.01 g, MB adsorption reached 87.79%. Ayalew *et al.* (2020) showed the efficacy of coffee husk-based biosorbent for MB dye removal. The adsorbent's specific surface area was measured at 28.54 m^2/g. At a pH of 5, with an initial dye concentration of 20 mg/L, an adsorbent dosage of 0.8 g/50 mL, a contact period of 50 min, and a temperature of 30°C on the activation surface of coffee husk, the maximum removal efficiency was obtained as 96.9%. With 6.82 mg/g at 30°C, the Langmuir model was determined to best fit the equilibrium data for MB adsorption. The pseudo-second-order model governs the adsorption procedure. MB was endothermically and spontaneously adsorbed onto the activated coffee husk, according to a thermodynamics analysis. The experimental results from the current study demonstrated that coffee husk is an effective biosorbent for cationic dye removal. Idibie (2021) prepared activated carbon from local papaya seeds and tested MB dye removal. The removal of MB from an aqueous solution was optimized, and its kinetics was studied. In order to achieve the best procedure, the impacts of operating variables such as time, adsorbent dosage, pH, and temperature were examined. The results showed that an ideal equilibrium adsorption of MB from an initial MB concentration of 350 mg/L, corresponding to a percentage removal of 91.46%, required an ideal time (50 mins), an ideal adsorbent dose (200 mg), an ideal pH (6.5), and an ideal temperature (60°C). Jawad *et al.* (2021) developed *Punica granatum* peel biochar (PBC) and environment-friendly biochar *via* a relatively fast acid-activation process with the great potential to be a promising adsorbent for the removal of MB dye. The PBC dosage of 0.18 g, temperature of 49°C, pH of 9.7, and time of 4.3 h were shown to be the best numerical adsorption conditions for achieving the maximum MB dye elimination (93.9%). The Temkin and Langmuir isotherm models performed quite well in fitting the equilibrium data. Using the Langmuir isotherm model, the SPPBC's greatest measured adsorption capacity for MB dye was 161.9 mg/g. This study demonstrates the feasibility of turning pomegranate peel lignocellulose into a sustainable material. Jabar *et al.* (2022) employed African almond leaves (ALs) to create a sustainable and environmentally friendly adsorbent. Equal weight was given to the kinetic and equilibrium adsorption isotherm models. Mesoporous PALB, which has a surface area of 816 m2/g and a radius of 1 nm, has an outstanding removal efficiency because it was able to remove >98% of the MB dye from an aqueous solution at 303 K, pH 8, initial dye concentration of 50 mg/L, adsorption dose of 0.2 g/L, and contact period of 30 min. Freundlich and pseudo-first-order (PFO) kinetic models provided the best fits for the adsorption isotherm and rate equation. Dural *et al.* (2011) used (*Posidonia ocionia* L.)-based activated carbon for the removal of MB dye removal under various starting concentrations, carbon dosages, pHs, and temperature settings, the adsorption potential of POAC for the removal of MB from aqueous solutions was also explored, and the best experimental conditions were identified. The samples showed reasonably quick kinetics that followed the second-order rate equation and reached equilibrium in around 60 min, whereas adsorption was unaffected by the solution's pH. The findings demonstrated that the suggested precursor led to AC with an improved MB sorption capacity, measuring 285.7 mg/g at 318 K. In conclusion, decaying leaves from the plant *P. oceanica* (L.) can be used as a raw material to create high-quality activated carbon.

3.2 *Effect of different parameters on MB adsorption*

The adsorption capacity of nanofillers and agricultural wastes for MB from wastewater can be affected by various parameters, including pH, contact time, initial dye concentration, and temperature. pH is a particularly important parameter, as it can affect the surface charge of both the dye molecules and the nanofillers, which can in turn affect the adsorption process. One of the most important factors affecting an adsorbent's ability to absorb material is pH. The degree of adsorptive molecule ionization and the surface properties of the adsorbent are influenced by the pH of the solution. The effect of the pH value on different agricultural waste adsorbents is shown in Figure 3. A study by Das *et al.* (2021) investigated the effect of pH on the adsorption capacity of mesoporous silica nanoparticles for the removal of MB from wastewater. The study found that the adsorption capacity of the nanoparticles increased with increasing pH, with a maximum adsorption capacity of 68.58 mg/g at a pH of 9. The increase in adsorption capacity was attributed to the increased electrostatic attraction

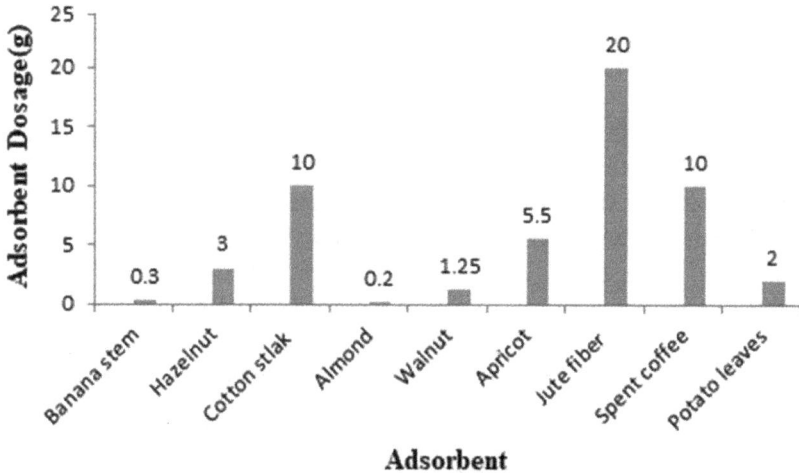

Figure 3.   pH value at which maximum MB adsorption was achieved.

between the positively charged nanoparticles and the negatively charged MB molecules at higher pH values.

Shakoor *et al.* (2016) revealed the effect of pH on MB adsorption using Punica granatum peel. Results showed that at a pH of 2, MB is adsorbed at a rate of 54.9%. The removal efficiency rose with the pH of the solution, peaking at 83.5% at pH = 7. Above a pH of 7, there is, however, no additional rise in removal efficiency. At low pH levels, H+ ion absorption causes the adsorbent's surface to become positively charged, which repels the cationic MB and the positively charged adsorbent. Because MB is a cationic dye, the more strongly negatively charged activated carbon will be more effective at absorbing it in an alkaline environment. The positive charge at the dye solution's interface decreases as the pH rises, and a negative charge develops on the adsorbent surface as a result. This leads to a greater electrostatic attraction between the positively charged adsorbate and the negatively charged adsorbent, which increases MB adsorption (Kaewsarn *et al.*, 2008). By utilizing coffee husk adsorbent at pH levels of 1, 4, 5, 6, and 7, the impact of solution pH on adsorption was examined (Ayalew *et al.*, 2020). When the initial pH solution was raised from 1 to 5, the removal performance rose from 87% to 97% and achieved its maximum level at a pH of 5. The values of pHs of different adsorbents are shown in Figure 3.

Similarly, a study by Bora *et al.* (2017) investigated the effect of pH on the adsorption of MB by activated carbon nanoparticles. The study found that the maximum adsorption capacity was achieved at a pH of 7, with a lower adsorption capacity at both lower and higher pH values. The decrease in adsorption capacity at low pH values was attributed to the competition between hydrogen ions and the dye molecules for adsorption sites on the activated carbon nanoparticles, while the decrease in adsorption capacity at higher pH values was attributed to the decrease in the positive charge of the activated carbon nanoparticles.

MB adsorption is greatly influenced by the initial concentration (Al-Gheethi *et al.*, 2022). On activated coffee husk adsorbents, the impact of starting dye concentration at 20, 40, 60, 80, and 100 mg/L adsorption was assessed (Ayalew *et al.*, 2020). As the initial dye concentration rises, the fraction of MB molecules eliminated falls. This might be the result of the adsorbent's active sites becoming saturated and condensation polymerization forming during the adsorption process. The percentage of MB dye removed decreases from 95.01% to 82.73% at 30°C, pH 5, 0.8 g adsorbent dose, and 50 min of contact time. Jamion *et al.* (2017) investigate the effect of initial MB concentration on the rate of adsorption using activated carbon from tamarind seed. The concentration of the MB in the range of 30, 60, 90, 120, and 150 ppm was used to determine the impact of dye concentration. The outcomes of the MB's ability to adsorb on prepared activated carbon showed that the adsorption capacity increased in the MB concentrations from 30 to 120 ppm. The absorption percentage was increased from 77% to 87%. However, at 150 ppm, MB's adsorption capacity was reduced. Variations in the initial dye concentration have an effect on different adsorbents.

The contact time is an important parameter that affects the adsorption capacity of nanofillers as well as agricultural wastes for MB from wastewater. It is defined as the time for which either nanofillers or agricultural wastes are in contact with the dye solution, and it can influence the rate and extent of dye adsorption. To determine the rate of dye removal, the relationship between MB adsorption and contact time was also examined by Kumar *et al.* (2010) by using cashew nut shell. Adsorption was found to be more rapid in the first 60 min of contact time and then gradually decreased until equilibrium was reached in around 90 min. Adsorption rises with increasing contact duration. An increase in contact duration up to 120 min revealed that the CNSAC's MB elimination was only 0.8% greater than with 90 min of contact time. The amount of MB that natural JSP eliminated as a result of adsorbent dosage. A study by Rajeswari and Srivastava (2016) investigated the effect of contact time on the adsorption capacity of graphene oxide for the removal of MB from wastewater. The study found that the adsorption capacity increased with increasing contact time up to 60 minutes, after which it plateaued. The initial increase in adsorption capacity was attributed to the availability of more adsorption sites on the graphene oxide surface, while the plateauing was attributed to the saturation of these sites.

Different adsorbents with various contact times and equilibrium adsorption capacities are shown in Table 4.

Adsorbent dosage is an important factor that affects the efficiency of the adsorption process. The amount of adsorbent used in the process affects the surface area available for adsorption, the availability of active sites, and the contact time between the adsorbent and the adsorbate.

Several studies have investigated the effect of adsorbent dosage on the adsorption capacity of nanofillers on MB from wastewater. For instance, in a study by Khan *et al.* (2021), the adsorption capacity of montmorillonite nanofillers on MB increased with increasing adsorbent dosage.

Table 4.   Contact time and respective adsorption capacity of different bio sorbents.

Adsorbent (Agricultural wastes)	Contact time (minute)	Equilibrium adsorption capacity (mg/g)	Reference
Coconut coir dust	20	29.5	Etim *et al.*, 2016
Onion skin	150	250	Saka *et al.*, 2011
Mango leaf powder	120	156	Uddin *et al.*, 2017
Magnetic nanoparticles	40	143	Pacheco *et al.*, 2019
Rice husk	7	25.46	Quansah *et al.*, 2020
Sugarcane bagasse	30	84.74	El Messaoudi *et al.*, 2016
Acacia wood	180	210.21	Fathy *et al.*, 2013
Rice straw	120	32.6	Khodaie *et al.*, 2013
Corn husk	80	462.96	Tiwari *et al.*, 2015
Potato peel	30	40	Hameed *et al.*, 2009
Garlic peel	210	82.64	Saka *et al.*, 2011
Cashew nut	60	250	Saber-Samandari *et al.*, 2015
Neem bark	30	55	Azadeh *et al.*, 2015
Pineapple stem waste	330.05	119	Etim *et al.*, 2016
Carica papaya seeds	50	320.14	Idibie, 2021
Meranti sawdust	180	342.9	Ahmad *et al.*, 2021
palm kernel fiber	60	94.5	El-Sayed *et al.*, 2011
Guava seed	—	0.10	Elizalde *et al.*, 2009
Grass waste	70	457.64	Hameed *et al.*, 2009
Bambo	480	286.1	Bello *et al.*, 2010

The authors attributed this to the increased surface area and availability of active sites as the dosage increased. Similar study was cited by Liu *et al.* (2020).

Nipa *et al.* (2019) studied the effect of the adsorbent dosage of jute stick powder. With larger adsorbent doses, the decolorization efficiency improves. Improved surface area and subsequently accessible adsorptive sites for efficient adsorption may be the cause of the increased decolorization efficacy at large adsorbent dosages. Since further increases in adsorbent dosages did not improve decolorization efficiency, the optimal adsorbent dose was determined to be 20 g/L. The effectiveness of dye

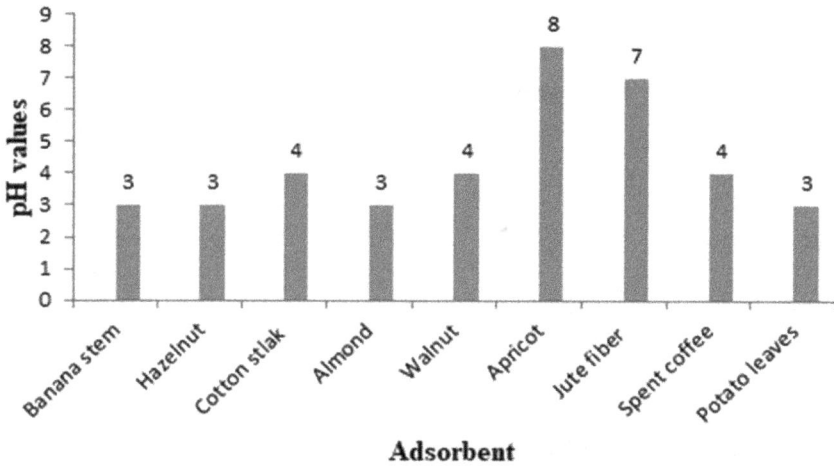

Figure 4. Adsorbent dosage at which maximum MB adsorption was achieved.

removal at the optimal adsorbent dosage was 81.72%. Ishak *et al.* (2022) demonstrated that the initial dosage of tamarind seed increased linearly with the amount of dye elimination. The rate of adsorption increases with an increasing adsorbent dosage. Different values of pH and adsorbent dosage on different agricultural wastes are observed in Figure 4. According to earlier studies, the dosage of the adsorbent boosted the effectiveness of color removal. This is because there are more binding sites available as the biosorbent dose rises.

3.3 *Adsorption isotherms models and kinetics*

An estimation of the adsorption capacity is provided by the adsorption isotherm, which is used to explain how the adsorbent interacts with the adsorbate (a monolayer or multilayer surface phase). In the literature, several isotherm models are presented in Table 5. The Freundlich isotherm model shows a variety of locations with different surface coverage (Figure 5(a)), whereas the monolayer adsorption that the Langmuir isotherm model predicts will occur on a surface with active accessible spots (Figure 5(b)). The adsorption isotherm is most frequently described using the Freundlich and Langmuir models. Dawood *et al.* (2016) used *Euclyptus sheathiana* bark. The Langmuir isotherm model provided the best representation of the equilibrium data and provided a monolayer effective adsorption capacity of biochar. The adsorption kinetics and

Table 5.   Adsorption capacity quantify by different isotherms.

Adsorbate	R^2 Langmuir isotherm	R^2 Freundlich isotherm	Maximum adsorption capacity	Reference
Rattan sawdust	0.999	0.76	294.14	Hameed *et al.*, 2007
Date palm seed	0.995	0.484	18.365	Abdus-Salam *et al.*, 2021
Areca catechu nut	0.9987	0.611	333.3	Joshi *et al.*, 2021
Yerba mate (Ilex paraguarensis)	0.912	0.889	59.6	Mazzeo *et al.*, 2020
Parkia speciosa pod	0.9966	0.9524	169.49	Aziz *et al.*, 2021
Pinang frond	0.991	0.979	384.62	Herawan *et al.*, 2020
Peanut hulls	0.9711	0.9339	265.41	Maind *et al.*, 2017

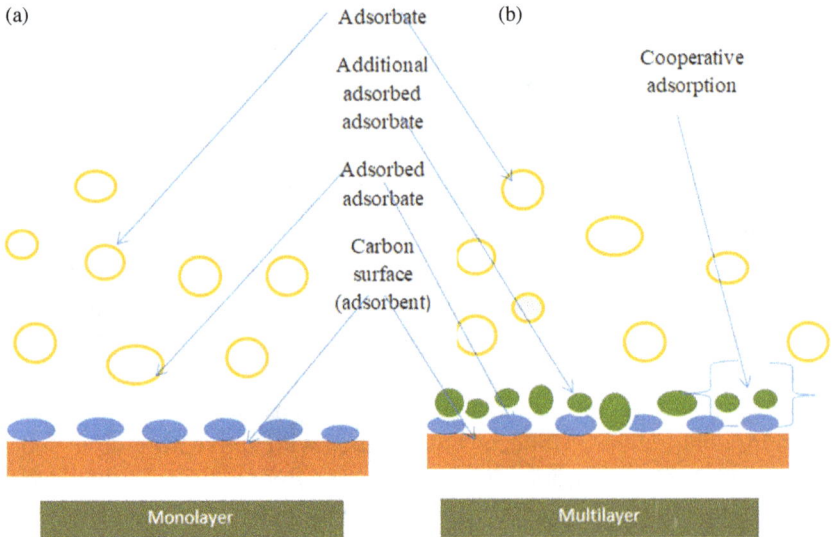

Figure 5.   (a) Monolayer and (b) multilayer adsorption.

mechanism of adsorption were further examined using PFO, pseudo-second-order, and intraparticle diffusion models that were fitted. Nguyen *et al.* (2021) conducted a study on the adsorption of MB onto *Acacia crassicarpa* material under optimal experimental conditions, including temperature, adsorbent dosage, and initial concentration of chemicals. The Langmuir isotherm model was observed to fit the adsorption data with an R^2 value of 0.9921, which is greater than the Freundlich value ($R^2 = 0.8922$), implying that the adsorption forms a monolayer. The maximum adsorption capacities predicted by the Langmuir isotherm were found to be 10.36 mg/g. Somaye *et al.* (2016) studied the adsorption of MB using rice agricultural waste. The isotherm study's results showed that the maximum adsorption capacity was 62.5 mg/g and that the adsorption data and the Langmuir isotherm model were in good agreement. In bamboo charcoal, studied by Zhu *et al.* (2009), the adsorption of MB was well described by the Langmuir isotherm, with R^2 values of 0.9995, 0.9996, and 0.9997 for the adsorption at 30, 40, and 50°C, respectively. The homogenous nature of the bamboo charcoal surface is indicated by the conformation of the experimental data matching the Langmuir isotherm model. The adsorption capacity increases with temperature, rising from 58.48 mg/g at 30°C to 64.10 mg/g at 40°C to 69.93 mg/g at 50°C.

Samarghandi *et al.* (2020) used cantaloupe peel for the removal of MB. The raw and modified cantaloupe peels absorbed 86.5 and 97% of the color, respectively, under ideal circumstances. Compared to the Freundlich, Temkin, Dubinin, and Radushkevich isotherms, the Langmuir isotherm provided a superior fit to adsorption isotherms, refer to Table 5. According to the pseudo-first-order, pseudo-second-order, Elovich, and intraparticle diffusion equations, the kinetic investigation of MB on raw and modified cantaloupe peel was carried out. The results gathered showed that, in addition to the interparticle diffusion model, the adsorption followed the pseudo-second-order rate. Mondal *et al.* (2012) utilize *Cocos nucifera* fruit shell for MB dye adsorption, and the maximum adsorption capacity was determined to be 20.74 mg/g at 40°C after the adsorption equilibrium data and the Langmuir model were successfully fitted. The dynamical data match the pseudo-second-order kinetic model effectively, according to kinetic analyses. Ghosh *et al.* (2020) utilized *Lathyrus sativus* husk for the removal of MB dye. The Langmuir isotherm model was well fitted for all the adsorbents at 313–323 K. The pseudo-second-order kinetic model provided the best fit. Saira *et al.* (2019) used *Malus domestica* sawdust as a low-cost biosorbent for MB removal. The experimental data fit well with the Langmuir model, with an R^2 value of 0.95. The PFO kinetic model is widely used in the adsorption process until equilibrium is attained (Idibie and Iyuke, 2008). The applicability of the PFO was tested for the adsorption of MB onto the locally produced CPSAC, as the linear regression correlation coefficient R^2 value obtained was 0.992.

3.5 *Mechanism of adsorption*

Adsorption occurs physically as a result of intermolecular gravity, namely van der Waals force or dispersion force. Because the surface particles of the adsorbent are not in the same environment as the particles inside the bulk, adsorption takes place. The interactions between the particles are all evenly balanced inside the adsorbent, but on the surface, where there aren't as many atoms or molecules of the same kind, the forces are unbalanced or still present. Adsorbate particles are drawn to their surface by these adsorbent forces. The surface diffusion process, which ascribes the gas or vapor along with the pores, occurs when the molecule adsorbate diffuses from the main body of the stream into the external surface of the adsorbent particle. Due to the presence of both surface diffusion and pore diffusion, it is known as mixed diffusion. In the adsorbent pores, where there is the most readily accessible surface area, the bulk adsorption takes place. The adsorbate then moves from the relatively small area of the external surface into the pores inside each adsorbent particle. The physical adsorption of porous adsorbents is influenced by the pore structure and specific surface area of such materials. The best adsorption capability can be obtained by having a higher microporous surface area (Al-Gheethi *et al.*, 2022).

4. Organic Nanoparticles

The word "nano" comes from a Greek word, and it means one billionth of a meter. The term "nanotechnology" was coined and used for the first time by Norio Taniguchi (1974) to illustrate the production and manufacturing of different materials at the nanometer scale. Nanotechnology is a dramatically growing area of research, with the potential to analyze and implement nanoscale materials in diverse areas of science and technology. It is a multidisciplinary scientific undertaking involving the creation and applications of materials, devices Figure 6, or systems that function at the nanoscale (Nakuma *et al.*, 2022).

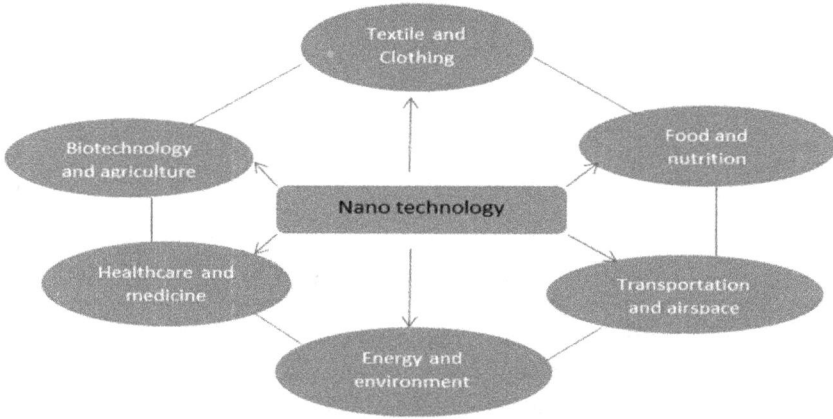

Figure 6.   Various applications of nanotechnology.

Organic nanoparticles, which include micelles, dendrimers, and ferritin, are biodegradable and non-toxic by nature (Nakuma *et al.*, 2022). Additionally, they might have hollow spheres with the same structure as micelles and liposomes, which are mostly used for medication delivery. Nanocapsules and polymeric nanoparticles are other names for them. They exhibit typical heat and light sensitivity. The most common forms are nanospheres and capsules. The outer boundaries of the spherical surface have absorbed chemicals, while the interior matrix portion of this nanoparticle (also known as a matrix particle) is solid (Mohammed *et al.*, 2017).

Fan *et al.* (2012) investigated the efficacy of magnetic β-cyclodextrin–chitosan (CDC) nanoparticles as nano-adsorbents for the removal of methyl blue. By attaching CDC to the surface of magnetite, a novel nanoadsorbent called CDC-modified Fe_3O_4 nanoparticles (CDCM) was created to remove MB from aqueous solutions. CDC was grafted onto Fe_3O_4 nanoparticles, according to the properties obtained from FTIR, SEM, and XRD. Because of the strong ability of CDCM to adsorb MB, which includes numerous hydroxyl, carboxyl, and amino groups, as well as the formation of an inclusion complex due to the -CD molecules through host–guest interactions, the grafted CDC on the Fe3O4 nanoparticles contributed to an improvement of the adsorption capacity. Temperature and pH were observed to affect the adsorption of MB onto CDCM.

It was discovered that the adsorption kinetics of MB follow a pseudo-second-order kinetic model and that adsorption equilibrium is reached in 50 min. The Langmuir isotherm model provides a good fit to the equilibrium data for MB adsorption. At 30°C, the maximum adsorption capacity of MB was calculated to be 2.78 g/g. The CDCM was steady and quickly bounced back. Additionally, after four uses, the adsorption capacity was around 90% of the initial saturation adsorption capacity.

5. Viewpoint and Upcoming Research

Due to their little or no cost, easy availability, and renewable nature, these waste products have been suggested more frequently than commercially activated carbon. Easily produced charcoal that can be utilized as activated carbon can be made from local agricultural waste. There is still a dearth of data, nevertheless, that offer a thorough analysis of sorbent comparisons. The current approach for the creation of activated carbon from farm wastes is to coat the adsorbents with nanoparticles to boost the removal process's effectiveness and generate highly applicable and reusable adsorbents that can be employed for several removal cycles. More research is needed in this area because the adsorption and desorption of MB utilizing adsorbents covered in nanoparticles have not been previously studied.

6. Conclusions

One of the cutting-edge techniques used to remove MB from textile effluent is adsorption, which utilizes adsorbents made of agricultural waste and nanofillers. Both nanofillers and activated carbon have advantages and limitations in textile wastewater treatment. Nanofillers have a high adsorption capacity and can remove pollutants at a very low concentration. However, they are expensive and can be difficult to separate from the wastewater once they have adsorbed pollutants. Activated carbon, on the other hand, is relatively inexpensive and easy to separate from the wastewater. However, it may not be as effective at removing pollutants at low concentrations as nanofillers. The initial dye concentration, pH, and dose of the adsorbents all affect the dye adsorption capabilities of agricultural waste adsorbents and nanofillers. The removal efficiency by adsorption ranges from 54.9% using *Punica garantum* peel to 98% using activated carbon coffee husk. All mechanisms and forms cannot be satisfied by a single model. Examples of models frequently used to explain isothermal adsorption in water and wastewater treatment applications are the Langmuir and Freundlich equations. Based on isotherm model investigations, MB adsorption occurs as mono- or multilayers.

References

Abbas, N. F. & Abbas, A. K. (2020). Novel complexes of thiobarbituric acid–azo dye: Structural, spectroscopic, biological activity and dying. *Biochemical and Cellular Archives, 20*(1), pp. 2419–2433.

Abdelwahab, O., El Nemr, A., El Sikaily, A., & Khaled, A. (2005). Use of rice husk for adsorption of direct dyes from aqueous solution: A case study of Direct F. Scarlet. *Egyptian Journal of Aquatic Research, 31*(1), pp. 1–11.

Abdus-Salam, N., Abiola, V. I.-U., & Fabian, A. U. (2021). Adsorptive removal of methylene blue from synthetic wastewater using date palm seeds, goethite and their composite. *Acta Scientifica Malaysia, 5*, 27–35.

Ahmad, A., Khan, N., Giri, B. S., Chowdhary, P., & Chaturvedi, P. (2020). Removal of methylene blue dye using rice husk, cow dung and sludge bio-char: Characterization, application, and kinetic studies. *Bioresource technology, 306*, p. 123202.

Ahmad, M. A., Ahmed, N. A. B., Adegoke, K. A., & Bello, O. S. (2021). Adsorptive potentials of lemongrass leaf for methylene blue dye removal. *Chemical Data Collections, 31*, p. 100578.

Al-Gheethi, A. A. *et al.* (2022). Sustainable approaches for removing Rhodamine B dye using agricultural waste adsorbents: A review. *Chemosphere, 287*, 132080.

Ayalew, A. A. & Aragaw, T. A. (2020). Utilization of treated coffee husk as low-cost bio-sorbent for adsorption of methylene blue. *Adsorption Science & Technology, 38*(5–6), pp. 205–222.

Azadeh, E., Seyed, F., & Ardovan, Y. (2015). Surfactant-modified wheat straw: Preparation, characterization and its application for MB adsorption from aqueous solution. *Journal of Chemical Engineering & Process Technology, 6*, 231.

Aziz, A., Hassan, H., Yahaya, N. E., Karim, J., & Ahmad, M. A. (2021). Methylene blue dye removal using *Parkia speciosa* pod based activated carbon. In *IOP Conference Series: Earth and Environmental Science* (Vol. 765, No. 1, p. 012104). IOP Publishing.

Babu, A. N., Reddy, D. S., Sharma, P., Kumar, G. S., Ravindhranath, K., & Mohan, G. K. (2019). Removal of hazardous indigo carmine dye from waste water using treated red mud. *Materials Today: Proceedings, 17*, pp. 198–208.

Bankole, P. O., Adedotun, A. A., & Sanjay, P. G. (2018). Enhanced decolorization and biodegradation of acid red 88 dye by newly isolated fungus, Achaetomium strumarium. *Journal of Environmental Chemical Engineering, 6*(2), 1589–1600.

Bayomie, O. S., Kandeel, H., Shoeib, T., Yang, H., Youssef, N., & El-Sayed, M. M. (2020). Novel approach for effective removal of methylene blue dye from water using fava bean peel waste. *Scientific Reports, 10*(1), p. 7824.

Bello, O. S., Adelaide, O. M., Hammed, M. A., & Popoola, O. A. M. (2010). Kinetic and equilibrium studies of methylene blue removal from aqueous solution by adsorption on treated sawdust. *Macedonian Journal of Chemistry and Chemical Engineering, 29*(1), pp. 77–85.

Bharti, V., Vikrant, K., Goswami, M., Tiwari, H., Sonwani, R. K., Lee, J., Tsang, D. C., Kim, K. H., Saeed, M., Kumar, S., & Rai, B. N. (2019). Biodegradation of methylene blue dye in a batch and continuous mode using biochar as packing media. *Environmental Research, 171*, pp. 356–364.

Birniwa, A. H., Abubakar, A. S., Mahmud, H. N. M. E., Kutty, S. R. M., Jagaba, A. H., Abdullahi, S. S. A., & Zango, Z. U. (2022). Application of agricultural wastes for cationic dyes removal from wastewater. In *Textile Wastewater Treatment: Sustainable Bio-nano Materials and Macromolecules*, Volume 1 (pp. 239–274). Singapore: Springer Nature Singapore.

Bora, H., Deka, R. C., & Saikia, J. P. (2017). Effect of pH on the adsorption of methylene blue dye from aqueous solution by activated carbon nanoparticles. *Journal of Environmental Chemical Engineering, 5*(5), 4728–4735.

Bosoy, D. *et al.* (2008). Utilization of methylene blue in the setting of hypotension associated with concurrent renal and hepatic failure: A concise review. *Opus, 12*, 21–29.

Choi, H.-J. & Sung-Whan, Y. (2019). Biosorption of methylene blue from aqueous solution by agricultural bioadsorbent corncob. *Environmental Engineering Research, 24*(1), 99–106.

Crini, G. & Eric, L. (2019). Advantages and disadvantages of techniques used for wastewater treatment. *Environmental Chemistry Letters, 17*(1), 145–155.

Da Silva, V. L., Kovaleski, J. L., Pagani, R. N., Silva, J. D. M., & Corsi, A. (2020). Implementation of Industry 4.0 concept in companies: Empirical evidences. *International Journal of Computer Integrated Manufacturing, 33*(4), pp. 325–342.

Dahlan, N. A., Lee, L., Pushpamalar, J. *et al.* (2019). Adsorption of methylene blue onto carboxymethyl sago pulp-immobilized sago waste hydrogel beads. *International Journal of Environmental Science and Technology, 16*, 2047–2058.

Das, P., Bora, U., & Dutta, J. (2021). Mesoporous silica nanoparticles: A pH responsive adsorbent for the removal of methylene blue from aqueous solution. *Journal of Environmental Chemical Engineering, 9*(1), 105169.

Dbik, A., Bentahar, S., El Khomri, M., El Messaoudi, N., & Lacherai, A. (2020). Adsorption of Congo red dye from aqueous solutions using tunics of the corm of the saffron. *Materials Today: Proceedings, 22*, pp. 134–139.

De, O., Rodrigues, G. *et al.* (2011). Electrocatalytic properties of Ti-supported Pt for decolorizing and removing dye from synthetic textile wastewaters. *Chemical Engineering Journal, 168*(1), 208–214.

Dragoi, E.-N., & Yasser, V. (2021). Modeling of mass transfer in vacuum membrane distillation process for radioactive wastewater treatment using artificial neural networks. *Toxin Reviews, 40*(4), 1526–1535.

Dural, M. U., Cavas, L., Papageorgiou, S. K., & Katsaros, F. K. (2011). Methylene blue adsorption on activated carbon prepared from *Posidonia oceanica* (L.) dead leaves: Kinetics and equilibrium studies. *Chemical Engineering Journal*, *168*(1), pp. 77–85.

El Farissi, H., Lakhmiri, R., Albourine, A., Safi, M., & Cherkaoui, O. (2021). Adsorption study of charcoal of cistus ladaniferus shell modified by H_3PO_4 and NaOH used as a low-cost adsorbent for the removal of toxic reactive red 23 dye: Kinetics and thermodynamics. *Materials Today: Proceedings*, *43*, pp. 1740–1748.

El Messaoudi, N., El Khomri, M., Bentahar, S., Dbik, A., Lacherai, A., & Bakiz, B. (2016). Evaluation of performance of chemically treated date stones: Application for the removal of cationic dyes from aqueous solutions. *Journal of the Taiwan Institute of Chemical Engineers*, *67*, pp. 244–253.

Elizalde-González, M. P., & Virginia, H.-M. (2009). Guava seed as an adsorbent and as a precursor of carbon for the adsorption of acid dyes. *Bioresource Technology*, *100*(7), 2111–2117.

El-Sayed, G. O. (2011). Removal of methylene blue and crystal violet from aqueous solutions by palm kernel fiber. *Desalination*, *272*(1–3), 225–232.

Etim, U. J., Umoren, S. A., & Eduok, U. M. (2016). Coconut coir dust as a low cost adsorbent for the removal of cationic dye from aqueous solution. *Journal of Saudi Chemical Society*, *20*, S67–S76.

Fan, L., Zhang, Y., Luo, C., Lu, F., Qiu, H., & Sun, M. (2012). Synthesis and characterization of magnetic β-cyclodextrin–chitosan nanoparticles as nano-adsorbents for removal of methyl blue. *International Journal of Biological Macromolecules*, *50*(2), pp. 444–450.

Fathy, N. A., Ola, I. E.-S., & Laila, B. K. (2013). Effectiveness of alkali-acid treatment in enhancement the adsorption capacity for rice straw: The removal of methylene blue dye. *International Scholarly Research Notices*, *2013*.

Feng, Y., Zhou, H., Liu, G., Qiao, J., Wang, J., Lu, H., Yang, L., & Wu, Y. (2012). Methylene blue adsorption onto swede rape straw (*Brassica napus* L.) modified by tartaric acid: Equilibrium, kinetic and adsorption mechanisms. *Bioresource Technology*, *125*, pp. 138–144.

Geer, S., Bernhardt-Barry, M. L., Garboczi, E. J., Whiting, J., & Donmez, A. (2018). A more efficient method for calibrating discrete element method parameters for simulations of metallic powder used in additive manufacturing. *Granular Matter*, *20*, pp. 1–17.

Gezer, B. & Yusuf, E. (2018). Adsorption behavior of methylene blue dye using carob powder as eco-friendly new adsorbent for cleaning wastewater: Optimization by response surface methodology. *Erzincan University Journal of Science and Technology*, *11*(2), 306–320.

Ghoreishi, S. M. & Haghighi, R. (2003). Chemical catalytic reaction and biological oxidation for treatment of non-biodegradable textile effluent. *Chemical Engineering Journal*, *95*(1–3), 163–169.

Ghosh, I., Kar, S., Chatterjee, T., Bar, N., & Das, S. K. (2021). Removal of methylene blue from aqueous solution using Lathyrus sativus husk: Adsorption study, MPR and ANN modelling. *Process Safety and Environmental Protection, 149*, pp. 345–361.

Gohr, M. S., Abd-Elhamid, A. I., El-Shanshory, A. A., & Soliman, H. M. (2022). Adsorption of cationic dyes onto chemically modified activated carbon: Kinetics and thermodynamic study. *Journal of Molecular Liquids, 346*, p. 118227.

Grayling, M. & Deakin, C. D. (2003). Methylene blue during cardiopulmonary bypass to treat refractory hypotension in septic endocarditis. *The Journal of Thoracic and Cardiovascular Surgery, 125*(2), 426–427.

Güler, M., Seda, Ç., & Deniz, B. (2021). Cinnamon bark as low-cost and eco-friendly adsorbent for the removal of indigo carmine and malachite green dyestuffs. *International Journal of Environmental Analytical Chemistry, 101*(6), 735–757.

Gupta, H. (2016). Photocatalytic degradation of phenanthrene in the presence of akaganeite nano-rods and the identification of degradation products. *RSC Advances, 6*(114), pp. 112721–112727.

Hameed, B. H. (2009). Grass waste: A novel sorbent for the removal of basic dye from aqueous solution. *Journal of Hazardous Materials, 166*(1), 233–238.

Hameed, B. H., Ahmad, A. L., & Latiff, K. N. A. (2007). Adsorption of basic dye (methylene blue) onto activated carbon prepared from rattan sawdust. *Dyes and Pigments, 75*(1), 143–149.

Hameed, B. H. & Ahmad, A. A. (2009). Batch adsorption of methylene blue from aqueous solution by garlic peel, an agricultural waste biomass. *Journal of Hazardous Materials, 164*(2–3), 870–875.

Hashem, A. H., Ebrahim, S., & Mohamed, S. H. (2020). Green and ecofriendly bio-removal of methylene blue dye from aqueous solution using biologically activated banana peel waste. *Sustainable Chemistry and Pharmacy, 18*, 100333.

Hassanein, T. F. & Koumanova, B. (2010). Evaluation of ad-sorption potential of the agricultural waste wheat straw for Basic Yellow 21. *Journal of the university of Chemical Technology and Metallurgy, 45*(4), 407–414.

Herawan, Safarudin, G., & Mohd, A. A. (2020). Study on adsorption of methylene blue on activated carbon from pinang frond using an experimental design to determine the optimum operating parameters. *IOP Conference Series: Earth and Environmental Science, 426*(1). IOP Publishing.

Hethnawi, A., Nassar, N. N., Manasrah, A. D., & Vitale, G. (2017). Polyethylenimine-functionalized pyroxene nanoparticles embedded on Diatomite for adsorptive removal of dye from textile wastewater in a fixed-bed column. *Chemical Engineering Journal, 320*, pp. 389–404.

Ho, Y.-S. & McKay, G. (2003). Sorption of dyes and copper ions onto biosorbents. *Process Biochemistry, 38*(7), 1047–1061.

Holkar, C. R., Jadhav, A. J., Pinjari, D. V., Mahamuni, N. M., & Pandit, A. B. (2016). A critical review on textile wastewater treatments: Possible approaches. *Journal of Environmental Management, 182*, pp. 351–366.

Idibie, A. C. & Iyuke, E. S. (2008). Thermodynamics study of linamarin sorption during its isolation from cassava. *Colloids and Surfaces A: Physicochemical and Engineering Aspects, 326*(1–2), pp. 18–22.

Idibie, C. A. (2021). Process optimization and kinetics study of locally produced activated carbon from Carica Papaya seeds on methylene blue adsorption. *FUPRE Journal of Scientific and Industrial Research, 5*(2), 21–29.

International Agency for Research on Cancer (1987). Overall evaluations of carcinogenicity: An updating of IARC monographs vol. 1 to 42. IARC monographs on the evaluation of the carcinogenic risk of chemicals to humans, *Suppl 7*(7), 1–440.

Ishak, Z. & Dilip, K. (2022). Adsorption of methylene blue and reactive black 5 by activated carbon derived from tamarind seeds. *Tropical Aquatic and Soil Pollution*, 2(1), 1–12.

Isik, Z., Arikan, E. B., Bouras, H. D., & Dizge, N. (2019). Bioactive ultrafiltration membrane manufactured from Aspergillus carbonarius M333 filamentous fungi for treatment of real textile wastewater. *Bioresource Technology Reports*, 5, pp. 212–219.

Jabar, J. M., Odusote, Y. A., Ayinde, Y. T., & Yılmaz, M. (2022). African almond (*Terminalia catappa* L) leaves biochar prepared through pyrolysis using H3PO4 as chemical activator for sequestration of methylene blue dye. *Results in Engineering, 14*, p. 100385.

Jamion, N. A. & Hashim, I. N. (2017). Preparation of activated carbon from tamarind seeds and methylene blue (MB) removal. *Journal of Fundamental and Applied Sciences, 9*(6S), 102–114.

Jawad, A. H., Rashid, R. A., Ishak, M. A. M., & Ismail, K. (2018). Adsorptive removal of methylene blue by chemically treated cellulosic waste banana (Musa sapientum) peels. *Journal of Taibah University for Science, 12*(6), pp. 809–819.

Joshi, S., Shrestha, R. G., Pradhananga, R. R., Ariga, K., & Shrestha, L. K. (2021). High surface area nanoporous activated carbons materials from Areca catechu nut with excellent iodine and methylene blue adsorption. *C, 8*(1), p. 2.

Kaewsarn, P., Wanna, S., & Surachai, W. (2008). Dried biosorbent derived from banana peel: A potential biosorbent for removal of cadmium ions from aqueous solution. *Proceedings of the 18th Thailand Chemical Engineering and Applied Chemistry.*

Khodaie, M. *et al.* (2013). Removal of methylene blue from wastewater by adsorption onto $ZnCl_2$ activated corn husk carbon equilibrium studies. *Journal of Chemistry, 2013.*

Kumar, P. S., Ramalingam, S., Senthamarai, C., Niranjanaa, M., Vijayalakshmi, P., & Sivanesan, S. (2010). Adsorption of dye from aqueous solution by cashew nut shell: Studies on equilibrium isotherm, kinetics and thermodynamics of interactions. *Desalination*, 261(1–2), 52–60.

Kuntari, K. & Febi, I. F. (2018). Utilization of bamboo leaves wastes for methylene blue dye adsorption. *AIP Conference Proceedings*, 2026(1). AIP Publishing LLC.

Lewoyehu, M. (2021). Comprehensive review on synthesis and application of activated carbon from agricultural residues for the remediation of venomous pollutants in wastewater. *Journal of Analytical and Applied Pyrolysis*, 159, 105279.

Li, G. *et al.* (2016). Effect of a magnetic field on the adsorptive removal of methylene blue onto wheat straw biochar. *Bioresource Technology*, 206, 16–22.

Maind, S. D., Dandekar, S., Lotlikar, O. A., Salunke, S., & Rathod, S. V. (2017). Effectiveness of Peanut Hulls (*Archis hypogaea* Linn.) for Removal of Cationic Dye, Methylene Blue from Aqueous Solution.

Manavi, N., Amir, S. K., & Babak, B. (2017). The development of aerobic granules from conventional activated sludge under anaerobic-aerobic cycles and their adaptation for treatment of dyeing wastewater. *Chemical Engineering Journal*, 312, 375–384.

Mazzeo, L. *et al.* (2020). Yerba Mate (*Ilex paragurensis*) as bio-adsorbent for the removal of methylene blue, remazol brilliant blue and chromium hexavalent: Thermodynamic and kinetic studies. *Water*, 12(7), 2016.

McMullan, G., Meehan, C., Conneely, A., Kirby, N., Robinson, T., Nigam, P., Banat, I., Marchant, R., & Smyth, W. F. (2001). Microbial decolourisation and degradation of textile dyes. *Applied Microbiology and Biotechnology*, 56, pp. 81–87.

Mohammed, M., Shitu, A., & Ibrahim, A. (2014). Removal of methylene blue using low cost adsorbent: A review. *Research Journal of Chemical Science*, 2231, 606X.

Mohammed, L., Gomaa, H. G., Ragab, D., & Zhu, J. (2017). Magnetic nanoparticles for environmental and biomedical applications: A review. *Particuology*, 30, pp. 1–14.

Mondal, N. S., Mondal, P., Roy, P. K., Mazumdar, A., & Majumder, A. (2012). Adsorption kinetics and isotherm modeling of arsenic removal from groundwater using electrocoagulation. *Journal of the Iranian Chemical Society*, pp. 1–12.

Mondal, P. K., Rais, A., & Rajeev, K. (2014). Adsorptive removal of hazardous methylene blue by fruit shell of Cocos nucifera. *Environmental Engineering & Management Journal*, 13(2).

Mustafa, H. M. & Gasim, H. (2021). Recent studies on applications of aquatic weed plants in phytoremediation of wastewater: A review article. *Ain Shams Engineering Journal*, 12(1), 355–365.

Nakum, J. & Bhattacharya, D. (2022). Various green nanomaterials used for wastewater and soil treatment: A mini-review. *Frontiers in Environmental Science*, 9, p. 724–814.

Naza, S., Alamb, S., Rehanc, K., & Sultanad, S. (2019). Adsorptive removal of new methylene blue from water by treated Malus domestica sawdust as a low cost biosorbent-equilibrium, kinetics and thermodynamic studies. *Desalination and Water Treatment, 166*, pp. 72–82.

Neme, I., Girma, G., & Chandran, M. (2022). Preparation and characterization of activated carbon from castor seed hull by chemical activation with H_3PO_4. *Results in Materials, 15*, 100304.

Nguyen, T. N., Dang, K. Q., & Nguyen, D. T. (2021). Adsorptive removal of methyl orange and methylene blue from aqueous solutions with Acacia crassicarpa activated carbon. *Vietnam Journal of Science, Technology and Engineering, 63*(4), pp. 23–27.

Nipa, S. T., Rahman, M. W., Saha, R., Hasan, M. M., & Deb, A. (2019). Jute stick powder as a potential low-cost adsorbent to uptake methylene blue from dye enriched wastewater. *Desalin Water Treat, 153*, pp. 279–287.

Noman, E., Al-Gheethi, A., Mohamed, R. M. S. R., & Talip, B. A. (2019). Mycoremediation of xenobiotic organic compounds for a sustainable environment: A critical review. *Topics in Current Chemistry, 377*, pp. 1–41.

Odoemelam, S. A., Ugbomma, N. E., & Nnabuk, O. E. (2018). Experimental and computational chemistry studies on the removal of methylene blue and malachite green dyes from aqueous solution by neem (*Azadirachta indica*) leaves. *Journal of Taibah University for Science, 12*(3), 255–265.

Oladoye, P. O., Ajiboye, T. O., Omotola, E. O., & Oyewola, O. J. (2022). Methylene blue dye: Toxicity and potential elimination technology from wastewater. *Results in Engineering, 16*, p. 100678.

Oyarce, E., Pizarro, G. D. C., Oyarzún, D. P., Martin-Trasanco, R., & Sánchez, J. (2020). Adsorption of methylene blue in aqueous solution using hydrogels based on 2-hydroxyethyl methacrylate copolymerized with itaconic acid or acrylic acid. *Materials Today Communications, 25*, p. 101324.

Pacheco, L. G., Nunes, V. H., Tonholo, J., da Silva, M. G. C., & Alves, C. R. (2019). Adsorption of methylene blue onto magnetic nanoparticles coated with chitosan. *Journal of Environmental Chemical Engineering, 7*(4), 103144.

Pavan, F. A., Ana, C. M., & Yoshitaka, G. (2008). Removal of methylene blue dye from aqueous solutions by adsorption using yellow passion fruit peel as adsorbent. *Bioresource Technology, 99*(8), 3162–3165.

Ponraj, C., Vinitha, G., & Joseph, D. (2017). A review on the visible light active BiFeO3 nanostructures as suitable photocatalyst in the degradation of different textile dyes. *Environmental Nanotechnology, Monitoring & Management, 7*, 110–120.

Praveen, S., Gokulan, R., Pushpa, T. B., & Jegan, J. (2021). Techno-economic feasibility of biochar as biosorbent for basic dye sequestration. *Journal of the Indian Chemical Society, 98*(8), p. 100107.

Pronk, W., Ding, A., Morgenroth, E., Derlon, N., Desmond, P., Burkhardt, M., Wu, B., & Fane, A. G. (2019). Gravity-driven membrane filtration for water and wastewater treatment: A review. *Water Research, 149*, pp. 553–565.

Qu, F., Li, H., Liu, X., Zhang, Y., Xu, L., & Wang, Y. (2019). Comparison of the adsorption capacity of activated carbon and graphene oxide for methylene blue. *Journal of Environmental Chemical Engineering, 7*(1), 102948.

Quansah, J. O., Hlaing, T., Lyonga, F. N., Kyi, P. P., Hong, S. H., Lee, C. G., & Park, S. J. (2020). Nascent rice husk as an adsorbent for removing cationic dyes from textile wastewater. *Applied Sciences, 10*(10), p. 3437.

Rafatullah, M., Sulaiman, O., Hashim, R., & Ahmad, A. (2010). Adsorption of methylene blue on low-cost adsorbents: A review. *Journal of Hazardous Materials, 177*(1–3), pp. 70–80.

Rahimian, R. & Soroush, Z. (2020). A review of studies on the removal of methylene blue dye from industrial wastewater using activated carbon adsorbents made from almond bark. *Progress in Chemical and Biochemical Research, 3*(3), 251–268.

Rajeswari, V. & Srivastava, S. (2016). Graphene oxide as an efficient adsorbent for the removal of methylene blue dye: A study of the adsorption kinetics and thermodynamics. *Journal of Environmental Chemical Engineering, 4*(3), 2699–2709.

Remya, N. & Jih-Gaw, L. (2011). Current status of microwave application in wastewater treatment — A review. *Chemical Engineering Journal, 166*(3), 797–813.

Robinson, T., McMullan, G., Marchant, R., & Nigam, P. (2001). Remediation of dyes in textile effluent: A critical review on current treatment technologies with a proposed alternative. *Bioresource Technology, 77*(3), pp. 247–255.

Russo, V., Masiello, D., Trifuoggi, M., Di Serio, M., & Tesser, R. (2016). Design of an adsorption column for methylene blue abatement over silica: From batch to continuous modeling. *Chemical Engineering Journal, 302*, pp. 287–295.

Saber-Samandari, S. & Jalil, H. (2015). Onion membrane: An efficient adsorbent for decoloring of wastewater. *Journal of Environmental Health Science and Engineering, 13*(1), 1–12.

Saeed, A., Sharif, M., & Iqbal, M. (2010). Application potential of grapefruit peel as dye sorbent: Kinetics, equilibrium and mechanism of crystal violet adsorption. *Journal of Hazardous Materials, 179*(1–3), pp. 564–572.

Saka, C. & Ömer, S. (2011). Removal of methylene blue from aqueous solutions by using cold plasma-and formaldehyde-treated onion skins. *Coloration Technology, 127*(4), 246–255.

Samarghandi, M. R., Godini, K., Azarian, G., Ehsani, A. R., & Zolghadrnasab, H. (2020). Adsorptive removal of methylene blue in aqueous solutions through raw and modified cantaloupe peel wastes: Kinetic and isotherm study. *Avicenna Journal of Environmental Health Engineering, 7*(1), pp. 35–46.

Sara, D., Tushar, K. S., & Chi, P. (2016). Adsorption removal of Methylene Blue (MB) dye from aqueous solution by bio-char prepared from *Eucalyptus sheathiana* bark: Kinetic, equilibrium, mechanism, thermodynamic and process design. *Desalination and Water Treatment, 57*(59), 28964–28980.

Shakoor, S. & Abu, N. (2016). Removal of methylene blue dye from artificially contaminated water using citrus limetta peel waste as a very low cost adsorbent. *Journal of the Taiwan Institute of Chemical Engineers, 66*, 154–163.

Sharma, P., Aramide, F. O., & Kushal, Q. (2022). Development of green geoadsorbent pellets from low fire clay for possible use in methylene blue removal in aquaculture. *Materials Today: Proceedings, 49*, 1556–1565.

Sivarajasekar, N. & Baskar, R. (2015). Agriculture waste biomass valorisation for cationic dyes sequestration: A concise review. *Journal of Chemical and Pharmaceutical Research, 7*(9), pp. 737–748.

Sleiman, M., Vildozo, D., Ferronato, C., & Chovelon, J. M. (2007). Photocatalytic degradation of azo dye Metanil Yellow: Optimization and kinetic modeling using a chemometric approach. *Applied Catalysis B: Environmental, 77*(1–2), pp. 1–11.

Sohrabi, H. & Elham, A. (2016). Adsorption equilibrium, kinetics, and thermodynamics assessment of the removal of the reactive red 141 dye using sesame waste. *Desalination and Water Treatment, 57*(38), 18087–18098.

Somaye, M., Hamedreza, J., Maryam, G., Tawfik, A. S., & Vinod, K. G. (2016). Microwave-induced H_2SO_4 activation of activated carbon derived from rice agricultural wastes for sorption of methylene blue from aqueous solution. *Desalination and Water Treatment, 57*(44), 21091–21104.

Srinivasan, S., Sadasivam, S. K., Gunalan, S., Shanmugam, G., & Kothandan, G., (2019). Application of docking and active site analysis for enzyme linked biodegradation of textile dyes. *Environmental Pollution, 248*, pp. 599–608.

Subramaniam, R. & Senthil, K. P. (2015). Novel adsorbent from agricultural waste (cashew NUT shell) for methylene blue dye removal: Optimization by response surface methodology. *Water Resources and Industry, 11*, 64–70.

Sulaiman, R., Adeyemi, I., Abraham, S. R., Hasan, S. W., & AlNashef, I. M. (2019). Liquid-liquid extraction of chlorophenols from wastewater using hydrophobic ionic liquids. *Journal of Molecular Liquids, 294*, p. 111680.

Tan, L., Xu, B., Hao, J., Wang, J., Shao, Y., & Mu, G. (2019). Biodegradation and detoxification of azo dyes by a newly isolated halotolerant yeast *Candida tropicalis* SYF-1. *Environmental Engineering Science, 36*(9), pp. 999–1010.

Tiwari, D. P., Singh, S. K., & Neetu, S. (2015). Sorption of methylene blue on treated agricultural adsorbents: Equilibrium and kinetic studies. *Applied Water Science, 5*(1), 81–88.

Uddin, M. T., Rahman, M. A., Rukanuzzaman, M., & Islam, M. A. (2017). A potential low cost adsorbent for the removal of cationic dyes from aqueous solutions. *Applied Water Science, 7*, pp. 2831–2842.

Uddin, M. K. & Abu, N. (2020). Walnut shell powder as a low-cost adsorbent for methylene blue dye: Isotherm, kinetics, thermodynamic, desorption and response surface methodology examinations. *Scientific Reports, 10*(1), 1–13.

Wang, X., Jiang, S., Tan, S., Wang, X., & Wang, H. (2018). Preparation and coagulation performance of hybrid coagulant polyacrylamide–polymeric aluminum ferric chloride. *Journal of Applied Polymer Science, 135*(23), p. 46355.

Wang, Y., Pan, J., Li, Y., Zhang, P., Zhang, X., Li, M., Zheng, H., Sun, Y., Wang, H., & Du, Q. (2021). Preparation and characterization of activated carbon from oil-palm fiber and its evaluation for methylene blue adsorption. *Materials and Technology, 55*(3), pp. 449–457.

Wu, Z., Wang, X., Yao, J., Zhan, S., Li, H., Zhang, J., & Qiu, Z. (2021). Synthesis of polyethyleneimine modified $CoFe_2O_4$-loaded porous biochar for selective adsorption properties towards dyes and exploration of interaction mechanisms. *Separation and Purification Technology, 277*, p. 119474.

Yang, Y., Liu, X., Li, W., Li, C., Yan, J., & Wang, S. (2020). Fabrication of a modified ZIF-8@PDMAEMA for efficient methylene blue removal. *Journal of Environmental Chemical Engineering, 8*(1), 103755.

Yaseen, D. A. & Scholz, M. (2019). Textile dye wastewater characteristics and constituents of synthetic effluents: A critical review. *International Journal of Environmental Science and Technology, 16*(2), 1193–1226.

Yu, H., Qu, F., Wu, Z., He, J., Rong, H., & Liang, H. (2020). Front-face fluorescence excitation-emission matrix (FF-EEM) for direct analysis of flocculated suspension without sample preparation in coagulation-ultrafiltration for wastewater reclamation. *Water Research, 187*, p. 116452.

Zaghbani, N., Amor, H., & Mahmoud, D. (2008). Removal of Safranin T from wastewater using micellar enhanced ultrafiltration. *Desalination, 222*(1–3), 348–356.

Zazou, H., Afanga, H., Akhouairi, S., Ouchtak, H., Addi, A. A., Akbour, R. A., Assabbane, A., Douch, J., Elmchaouri, A., Duplay, J., & Jada, A. (2019). Treatment of textile industry wastewater by electrocoagulation coupled with electrochemical advanced oxidation process. *Journal of Water Process Engineering, 28*, pp. 214–221.

Zhu, Y., Wang, D., Zhang, X., & Qin, H. (2009). Adsorption removal of methylene blue from aqueous solution by using bamboo charcoal. *Fresenius Environmental Bulletin, 18*(3), pp. 369–376.

Zietzschmann, F., Christian, S., & Martin, J. (2016). Granular activated carbon adsorption of organic micro-pollutants in drinking water and treated wastewater–aligning breakthrough curves and capacities. *Water Research, 92*, 180–187.

Chapter 7

Agricultural Floral Waste (AFW) and Scope of its Utilization in Agronomy

Sonali Ramgopal Mahule* and Roli Mishra†

*Amity School of Applied Science,
Amity University Mumbai, Mumbai-Pune Expressway Bhatan,
Somathne, Panvel 410206, Maharashtra*
*sonali.mahule@gmail.com
†rolimishraenviro@gmail.com*

Abstract

Solid waste management is increasingly challenging due to rapid urbanization, exponential population growth, and industrial expansion, making it one of the most pressing global issues today. A nation's overall solid waste output, which results from numerous agricultural, urban, cultural, social, and religious activities, causes environmental (air, water, and soil) pollution as well as health issues for its people. This in turn significantly damages the Earth's ecological environment. Agricultural Floral Waste (AFW) encompasses all types of solid organic waste generated from various human activities in fields and urban areas, making its management crucial. Every year, India generates about 500 million tons of farm waste, and Indian households waste 50 kg of food per capita. In financial terms, the harvest and post-harvest losses of major agricultural and allied produce amount to Rs. 92,651 crore yearly. In India, between 70 and 80% of solid waste is organic waste. Any organic waste, including floral debris, when dumped outdoors can lead to environmental issues since the decomposition process encourages the growth of germs. Appropriate handling of floral waste disposal will contribute to both the achievement of almost 60% of the UN's Sustainable Development Goals and a decrease in environmental pollution. Kenya and China have been turning flower waste into compost, but the United States and Europe are burning flower debris, which adds to air pollution. Efficient management can lessen environmental stress and yield a variety of beneficial resources such as compost, biogas, biochar, natural dyes, and perfumes. Therefore, to address the potential challenges associated with implementing the sustainable management of AFW in many parts of the world, including India, a thorough investigation is urgently needed.

Keywords: Agricultural waste, crop residue, circular economy, waste-to-wealth, composting, biopesticides.

1. Introduction

Flora refers to plant life in a particular region, period, or environment. Therefore, floral waste includes all the biodegradable waste generated from entire plant species, including leaves, flowers, vegetables, fruits, trimming remnants, and leftover crops after gleaning. Billions of tons of such agricultural floral waste (AFW) are generated worldwide every year from their respective sources, which accumulate and play a major role in polluting the ecological environment of the entire world (Agapkin *et al.*, 2022; Anand *et al.*, 2013; Tchonkouang *et al.*, 2023). Their production rate is directly proportional to the global population. Therefore, the disposal of all these types of AFW is a major global issue in this context (Agapkin *et al.*, 2022).

The utilization of AFW materials into respective value-added products is practiced around the world up to some extent by government waste management agencies and industries (Bala *et al.*, 2023b; Rai, 2023; Shu *et al.*, 2015; Zhu *et al.*, 2023). Obtaining a significant output is not so easy in this respect because it requires the collection and management of AFW from different locations. However, if they are not converted into value-added products, AFW will accumulate in their respective areas and become pollutants. Therefore, collecting bulk amounts of AFW material from different sources is the major challenge in this sector. Lately, numerous start-ups have been emerging in the agricultural waste management sectors and are producing biofertilizers, biofuels, and biopesticides and extracting bioactive substances from different kinds of AFW (Zhu *et al.*, 2023). Several methods are recommended for utilizing different types of AFW material processing (Sharma, 2021). Many of these methods are in practice and are based on the various physical and chemical properties of the different types of AFW. However, in India, farmers follow many conventional methods to clear out AFW from their fields, such as burning, dumping, and use of biogas plants. With the evolution of synthetic fertilizers and pesticides, as we pursue maximum production from the soil, such domestic methods of utilization of AFW became uncommon and finally extinct in the current era. Therefore, AFW becomes hazardous due to improper disposal in other places, such as water bodies and open lands (Waghmode *et al.*, 2018). Ultimately, this leads to harmful situations for plants and animals by causing pollution in the soil and air. In this chapter, we provide glimpses into various sources of AFW, its management techniques, and its utilization, especially in agriculture.

2. Sources of Agriculture Floral Waste

A nation's overall solid waste output results from numerous cultural, social, and religious activities. We can categorize these sources of AFW into five main sectors (Figure 1): religious places, agricultural and non-agricultural fields, food industries, Agricultural Produce Market Committee (APMC) markets, and urban wastes (domestic and hotel AFW).

2.1 *Religious places*

In Indian culture, flowers are most commonly associated with religious rituals, marriage ceremonies, and other celebrations and are presented as gifts. Due to a variety of religious practices, tons of flower waste frequently accumulate at places of worship, such as temples, mosques, churches, and gurudwaras. It is also produced in locations such as flower markets, wedding ceremonies, residential areas, community centers,

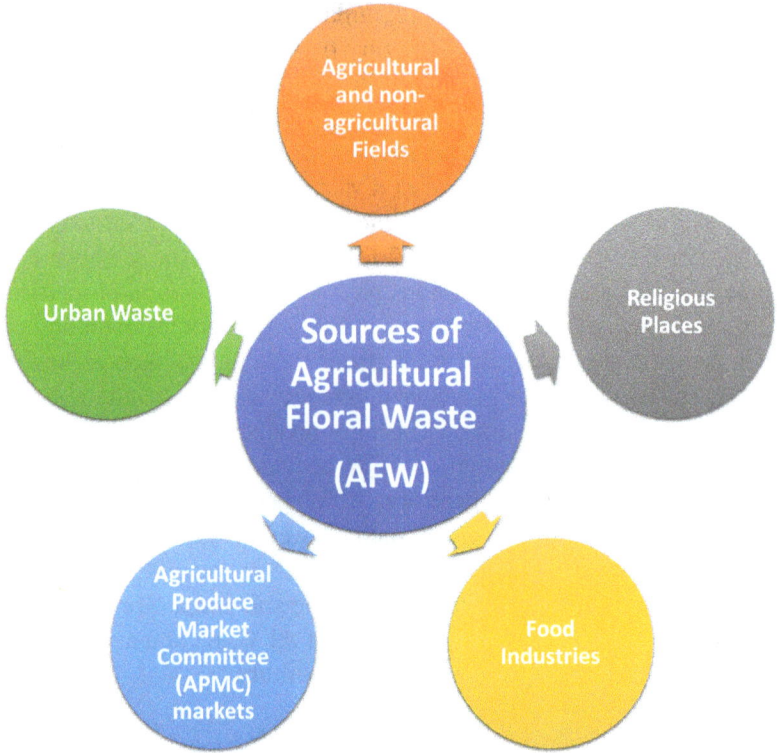

Figure 1.   AFW generating sectors.

hotels, and other traditional and sacred ceremonies (Waghmode *et al.*, 2018). The most common cut flowers produced in India are Anthurium, Carnation, Dutch Rose, Gerbera, Gladiolus, Orchids, and Tuberose Double; the most common loose flowers produced in India are Jasmine, Marigold, Rose Loose, and Tuberose Single (Indiastat, 2020). India's temples and banqueting rooms are adorned with enormous amounts of flowers, including marigold, chrysanthemum, rose, and jasmine. Beyond these locations, flower marketplaces also produce flower waste, particularly during periods of low market sales (Soundrya *et al.*, 2021). In temples, the most commonly offered flowers include roses, marigolds, jasmine, and hibiscus.

Temples are regarded as the dwelling places of the gods. Hindus visit temples as part of their custom to obtain the blessings of their gods before beginning any momentous occasion. Many deeply religious Hindus visit temples almost daily, and as a part of worship, flowers are essentially used. As a result, a huge amount of flower waste is generated from temples worldwide. According to government data, there are 108,000 mosques and temples listed, but the actual number may lie somewhere around 600,000. Every year, over 8 million tons of flowers are discarded into rivers, where harmful herbicides and insecticides used to cultivate the flowers mix with the waters (Mahindrakar, 2018). In India, many places of worship generate 20 tons of flower waste daily. Regrettably, the decomposition of these hallowed flowers in streams results in the death of fish, the destruction of the water body's fragile ecosystem, and severe contamination.

According to Abeliotis *et al.*, the European Union (EU), which generates about 44% of the world's total flower production, was predicted to have the largest share of global flower production in 2018, with the Netherlands alone estimated to contribute about one-third of this total (Abeliotis *et al.*, 2015). India is the nation with the most acreage dedicated to floriculture; in 2014–2015, it produced 477 lakh cut flowers and around 1,641 t of loose flowers (Horticulture–Statistical Yearbook India, 2016). In the same way, Arici *et al.* (2016) reported that Turkey produces about 50 t of tulip petal waste annually. Masure and Patil (2014) found that in developing nations such as India and Sri Lanka, 40% of the total flower production is unsold and wasted because of inadequate facilities like cold storage. According to a study, India generates roughly 4,738 t of floral waste every day (Sharma *et al.*, 2018). Varanasi and Surat, two of India's most important holy towns, produce roughly 10 and 1.5 t/d of flower waste, respectively (Sharma *et al.*, 2018). In the fiscal year 2023, it was projected that flower production in India would reach around 3 million metric tons. Compared to the previous fiscal year, there was a minor drop in output volume (Figure 2). On a typical day, at the Kashi Vishwanath Temple, about 2,500 kg of flowers and leaves are said to be offered each day, while during holidays and important occasions such as Shrawan month, the quantity increases significantly. Likewise, the Mumbai Municipal Corporation (BMC) gathers 60 t of flower offerings (*nirmalya*) from various Ganesh mandals at its 16 collecting locations in just four days following pooja activities, which is usually discarded into the sea. Every day, thousands of worshippers flock to the Shirdi Shri Saibaba temple to offer prayers and offerings to the god, resulting in the daily production of two to three tons of floral waste. Chilkur Balaji

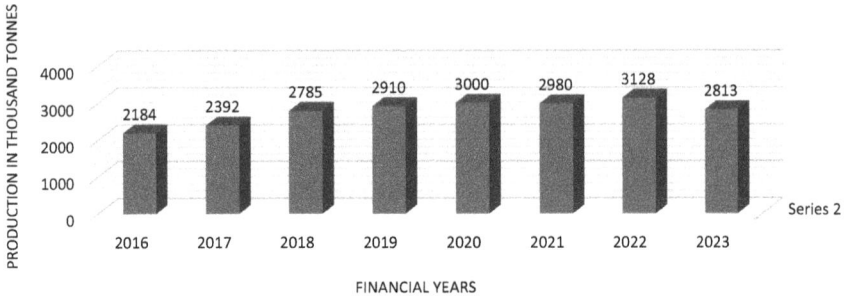

Figure 2.   Production of Volume of Flowers in India 2016–2023.

Temple, also known as Visa Balaji Temple, is an old Hindu temple dedicated to Lord Balaji and located close to Hyderabad, India, on the shores of Osman Sagar Lake. It produces about 60 kg of floral waste every day (Soundrya *et al.*, 2021). Similarly, every day, approximately 15–18 quintals of flowers are offered at Khwaja Moinuddin Chishti's Ajmer Sherif Dargah. After Andhra Pradesh, Karnataka, and Tamil Nadu, West Bengal ranks fourth in India for the highest flower waste generation (Waghmode *et al.*, 2016).

2.2 *Agricultural and non-agricultural fields*

Studies revealed that Ukraine produces 115 million tons (Mt) of agricultural plant waste each year while India produces, on average, 500 Mt of plant waste each year, according to the Ministry of New and Renewable Energy (MNRE) of India (Pryshliak & Tokarchuk, 2020; NPMCR, 2020). Statewide, in India, Uttar Pradesh produces the highest estimated amount of plant waste (60 Mt), while Maharashtra (46 Mt) and Punjab (51 Mt) are two other states that produce significant plant waste (Devi *et al.*, 2017). Paddy alone produces 34% of the waste from cereal crops such as wheat, maize, millets, and paddy (Lohan *et al.*, 2018), and approximately 92 metric tons of total plant waste are burned every year in India (Bhuvaneshwari *et al.*, 2019).

2.3 *Urban waste*

According to World Bank predictions, the global municipal solid waste generation in 2050 is expected to reach 3.4–109 tons (Kaza *et al.*, 2018).

Of this, kitchen trash and plant waste account for an average of 52.8–65.3% of the total (Ding *et al.*, 2021). Domestic plant waste comprises agricultural waste, rubbish from markets, garden waste, kitchen waste, and leaves (Exposto & Januraga, 2021).

2.4 *Agriculture Produce Marketing Committee (APMC) markets*

Agriculture Produce Marketing Committees (APMCs) generate a large quantity of plant waste from fruits and vegetables (Figure 3). Farmers do

Figure 3.   APMC market waste.

not grade or clean the produce before bringing it to the market. To increase profits, most products are sold by weight.

The waste remains uncollected and begins to decompose at the location because of the high volume and weight. There is a lot of agricultural trash in the market yard, including jowar and sorghum straw, dried leaves from trees and crops, the stalks of pigeon peas (*Cajanus cajan*), vegetable wastes, and parthenium weed (Mane *et al.*, 2012). According to data released by Soundrya *et al.* in 2021, each shop in Hyderabad's Gudimalkapur market sells an average of 526.5 kg of flowers each day, which results in an average of 51.75 kg of floral waste daily.

2.5 Food industries

In India, each year, over 18% of the plants produced from fruit and vegetable cultivation valued at Rs. 44,000 crore, are wasted (Rudra *et al.*, 2015). In terms of processed food production, consumption, and exports, India ranks fifth globally. According to Khedkar and Singh (2017, 2018), it is the world's top producer of fruits and vegetables, including bananas, guavas, papayas, mangoes, and ginger. It is also ranked second in the world for producing green peas, pumpkins, guards, and cauliflower. Poor harvesting techniques leave certain plant components behind during production or harvest, and fruits that do not satisfy quality requirements or are unprofitable to harvest are discarded, which creates a massive amount of plant waste.

3. Management of AFW

As there are no serious efforts from the municipality or local governing bodies to collect AFW, collection remains the major challenge. Collecting each type of AFW from all its sources, from the farm, market, religious places, and hotels to food processing plants, requires a planned management system to integrate this critical process into common practice.

There are certain methods for AFW management and its treatments. The purpose of these methods is to dispose of AFW using a particular technique, and all these processes generate simple organic matter (OM) as a product that can be further utilized in farming as manure to enrich the soil's nutrient content using natural sources (Singh *et al.*, 2022). The common disposal processes are briefly discussed in the following.

3.1 *Pyrolysis of AFW*

Pyrolysis is the process of heating the biomass to temperatures of 500°C or higher without air. This method is used to convert biomass into intermediate liquid materials that can be further converted into biofuels. Heating with no air decomposes the biomass into combustible gases and biochar. The combustible gases can be condensed into an organic liquid called bio-oil and another liquid. It also produces gaseous products, including syngas, which can be used as energy sources in process development. One such energy source is biomass fuels, which are used for energy production while reducing greenhouse gas emissions and other pollutants. This is essential for meeting the latest emission targets. The utilization of AFW as an energy resource increases the share of renewable energy sources. The most frequently applied thermochemical technologies for converting biomass into energy or chemicals are combustion, pyrolysis, gasification, and high-pressure liquefaction (Agata Mlonka-Mędrala *et al.*, 2021). In pyrolysis and gasification technologies, lower-quality fuels are acceptable, and pyrolysis serves as a means of releasing the stored energy within biomass by transforming it into other useful products. The main part of the process is the thermal cracking reaction, in which organic and inorganic gaseous compounds are released during sample heating. The initial products of pyrolysis are condensable gases and solid biochar. Further breakdown of the condensable gas produces the final gaseous products: carbon monoxide (CO), carbon dioxide (CO_2), hydrogen (H_2), methane (CH_4), some higher hydrocarbons such as ethyne (C_2H_2), ethylene (C_2H_4), and ethane (C_2H_6), together with liquid products — tars — and solid products in the form of char. The yields of pyrolysis products depend on the biomass composition, especially its hydrogen-to-carbon (H/C) ratio, and process parameters such as the heating rate, pressure, temperature, and residence time. Pyrolysis is a complex process and depends on many parameters. Several studies were dedicated to experimental studies of the waste biomass pyrolysis process, but only a few analyzed straw as a feedstock. Kinetic analysis and pyrolysis behavior of *Azadirachta indica* and *Phyllanthus emblica* kernels were analyzed by scientists from India (Mishra *et al.*,

2020). Wheat, flax, oat, and barley straws were considered feedstocks for the catalytic pyrolysis process; however, only one process temperature of 500°C was considered (Aqsha *et al.*, 2017). Similar studies of catalytic pyrolysis at 500°C on other materials, such as rice straw, sugar cane bagasse, ugu plant, and willow, were also carried out by Jaffar *et al.* (2020).

3.2 Composting of AFW

Composting is the conversion of complex organic waste into smaller organic molecules through the action of microorganisms such as bacteria and fungi under controlled conditions of temperature and moisture. In terms of both environmental and economic considerations, this is one of the best AFW treatment practices compared to other methods such as landfilling, pyrolysis, and incineration. Therefore, it makes a major contribution to recycling AFW. Composting requires a sequence of steps by which organic waste materials such as leaves, grass, fruits, and vegetables are acted upon by microorganisms in the presence of water and oxygen to produce humus. The humus is a natural fertilizer that is friendly to the environment and rich in fiber and inorganic nutrients such as phosphorus, potassium, and nitrogen. The composting process can vary with the method or equipment used. There are three types of composting, namely aerobic, anaerobic, and vermicomposting. The process can either be natural or accomplished with the aid of machines. The various benefits of using compost in crop production include enriching the soil by fixing valuable nitrogen, which helps the soil to retain moisture; suppressing plant diseases and pests; reducing chemical fertilizer use; encouraging the production of beneficial bacteria and fungi that help in breaking down OM to create a rich nutrient-filled material called humus, reducing methane emissions from landfills; and lowering our carbon footprint.

3.3 Landfills of AFW

Dumping, or landfilling, is the oldest and most common waste disposal method. The landfill is a well-designed pit lined at the bottom surface in which solid waste is filled. Designing a landfill requires expertise, and operating it requires skilled operators to ensure proper AFW management and functioning. This pit is used to bury trash in a way that prevents groundwater pollution, ensuring that it remains dry and does not come into contact with air. Landfill material (LM) is defined as anaerobically decomposed and excavated waste from a landfill site, containing organic as well as inorganic substances, which is used to fertilize agricultural soils. The use of LM in agricultural fields has numerous advantages. One of the advantages of LM is that it significantly improves crop yields because it contains OM and plant nutrients, such as nitrogen and phosphorus. However, there is also great concern that landfill-mining areas may contain considerable amounts of organic and inorganic pollutants, which can pollute agricultural soils over time. These pollutants include plastics, metals, glasses, fibers, various organic pollutants, and heavy metals. Although visible pollutants are perceived as a pollution problem, awareness about the potential risks of invisible pollutants such as heavy metals is lacking (Mehreteab Tesfai *et al.*, 2009).

3.4 *Sanitary landfills of AFW*

In this method, solid waste is disposed of on land while ensuring public health or safety is not compromised. This can be done by applying engineering principles to confine solid waste within a small piece of land and reducing the waste volume by periodically covering it with a layer of earth.

3.5 *Incineration*

Incineration is the process of treating waste through the combustion of organic substances present in waste materials. Industrial plants for waste incineration are commonly termed waste-to-energy facilities. Incineration and other high-temperature waste treatment systems are referred to as thermal treatment. The solid mass of the original waste is reduced by around 80–85%, while its volume is reduced by 95–96%. Incineration does not entirely replace landfilling; however, it considerably reduces the amount of waste to be disposed of. Incineration is becoming more and more important for the disposal of AFW in most developed countries because landfill sites are becoming more difficult to identify due to their high costs, diminishing land availability, more stringent regulations, and frequent public opposition to the siting of new landfills. Generally, incineration can effectively reduce municipal waste volume and may also provide usable energy (Bierman & Rosen, 1994). During incineration, N and C nutrients in municipal wastes are eliminated, non-volatile elements are concentrated in altered chemical forms, but P and K are enriched (Bierman & Rosen, 1994; Frossard *et al.*, 1996; Chang *et al.*, 1999; Enders & Spiegel, 1999; Linak *et al.*, 1999). A great deal of attention has been paid to P in municipal waste and its incineration ash. The amount of P in municipal waste is considerable, as reported by Sommers (1977) and Fine and Mingelgrin (1996), the P content in sewage sludges ranges from 15 to 30 g/kg on a dry weight basis. Most of the P in the sludge is in organic form, and 70–90% of the total P is extractable using ammonium citrate and acetic acid. Due to its relatively high P content, municipal waste incinerator ash may be used as a source of P for agricultural purposes. The application of waste materials to agricultural land presents an opportunity for the recovery of essential plant nutrients; many waste products contain concentrations of plant nutrient elements that are sufficient to produce an agriculturally significant growth response (Fuller & Warrick, 1985). It has been shown that incinerator ash can supply plants with available nutrients, act as an effective liming agent, and increase crop yields (Willems *et al.*, 1976; Mellbye *et al.*, 1982; Jakobsen & Willett, 1986; Bierman & Rosen, 1994). On the other hand, recycling waste materials in agricultural systems requires an evaluation of both the agronomic benefits and broader environmental consequences. The suitability of ash as a soil amendment is limited by its trace metal content because high concentrations of trace metals can cause phytotoxicity or even human and animal health problems (Fu-Shen Zhang *et al.*, 2001). However, incineration of agricultural waste in large quantities can have a negative impact on the environment and habitats through greenhouse gas emissions, unpleasant odors, and toxic liquids that may infiltrate water sources.

3.6 *The 3R approach reduce-reuse-recycle*

3R is the concept of efficiently managing waste. The top priority is to reduce waste generation, then reuse, and recycle the generated waste to give waste a second chance before disposing of it in landfills. This idea is best applicable to nonbiodegradable types of waste, such as electronics, metallic items, and plastics. However, for AFW, the reduction of waste can be achieved through practices such as not wasting food, reusing biodegradable items, and separating them for recycling. Reduce means to minimize the amount of waste we create. Reuse refers to using items more

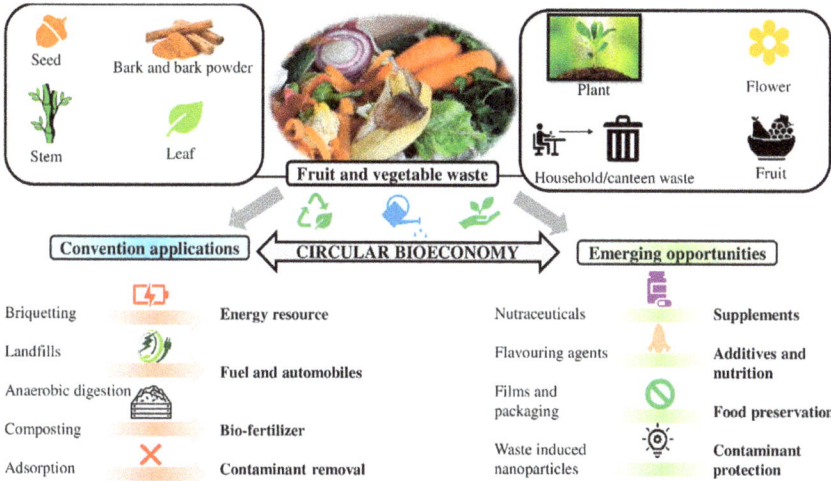

Figure 4.   Utilization of AFW to produce value-added products.

than once. Recycling means putting a product to new use instead of throwing it away.

3.7 *Advantages of the utility of AFW in agriculture*

From an ecological and sustainability point of view, the best utility of AFW is manure and pesticide production. This, in turn, reduces the quantity of synthetic fertilizers used in crop production and consequently soil quality is maintained over a longer period. The major sources of AFW are agricultural lands; therefore, if it is treated at the source, then that will also reduce the energy and capital required for the transportation of large amounts of organic wastes.

In the following, we describe a few popular and best-practice technologies employed for converting several types of AFW into green products (Figure 4) (Ganesh *et al.*, 2022).

4. AFW Utilization in Agriculture

The rate of degradation of AFW is a very slow process, meaning there is a need for a proper and eco-friendly treatment process. The waste generated can be used for making various bioactive compounds (Bala *et al.*,

2023a) and products such as biofuels (Kamusoko *et al.*, 2021), biofiber for making paper (Vigneswaran *et al.*, 2015), biopesticides, biomanure (NWOSIBE Patrick O and J, 2020), natural Holi colors, rose water essence, natural dyes, incense sticks, citric acid production, and antibiotics. The products and technologies for AFW utilization and the corresponding products derived from them are discussed as follows.

4.1 *Vermicompost production*

The production of manure through a biotechnological process is called vermicomposting. In this technique, certain species of earthworms are used for the conversion of AFW into an enriched product. Compost and vermicompost are the products of the aerobic composting process; the latter uses earthworms and is therefore obtained faster than compost (Sharmila Rani, 2020). The manure is rich in microbial activity and plant growth regulators and possesses pest repellent properties (Figure 5) (Rehman *et al.*, 2023). Thus, since ancient times, these worms have been called "*Shetkaryancha Mitra*" ("farmers' friend"), as they transform garbage into gold.

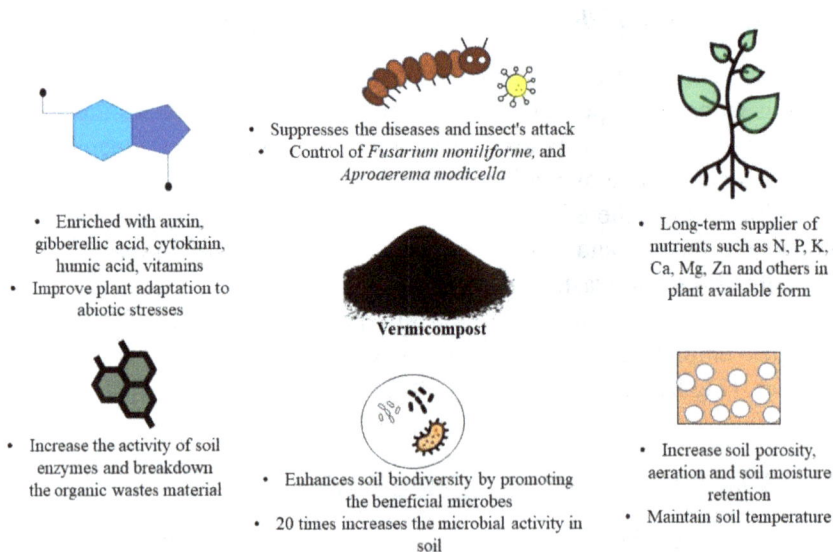

- Suppresses the diseases and insect's attack
- Control of *Fusarium moniliforme*, and *Aproaerema modicella*

- Enriched with auxin, gibberellic acid, cytokinin, humic acid, vitamins
- Improve plant adaptation to abiotic stresses

Vermicompost

- Long-term supplier of nutrients such as N, P, K, Ca, Mg, Zn and others in plant available form

- Increase the activity of soil enzymes and breakdown the organic wastes material

- Enhances soil biodiversity by promoting the beneficial microbes
- 20 times increases the microbial activity in soil

- Increase soil porosity, aeration and soil moisture retention
- Maintain soil temperature

Figure 5.   Vermicompost: Mode of action to enhance the plant growth.

Kohli and Hussain obtained 2.75 kg of vermicompost by using 5 kg each of flower waste and cow dung. Alternating layers of farm waste, floral waste, and cow dung were arranged, and around 200 earthworms were spread in partially digested material containing 60% moisture content. The vermiwash was diluted with water and sprayed on plants as an antifungal agent. The obtained vermicompost has a higher C/N ratio. Similarly, Sharma *et al.* produced vermicompost by mixing floral waste, cow dung, and sawdust. Vermicomposting of temple waste obtained from a Ganesh temple in Sangli, Maharashtra. They mixed the effluent from a biogas digester with temple waste and cattle dung. This was then allowed to decompose for a period of 30 days at 30°C. The vermicompost obtained was used as a fertilizer for five flowering plants, and the authors reported that good growth parameters were obtained in terms of height, flowering time, and the number of flowers produced when compared to the control setup, which was not treated with vermicompost. Therefore, vermicomposting of flower waste is an excellent and eco-friendly method of flower waste management. Jadhav *et al.* reportedly developed a microbial consortium for the effective degradation of flower waste generated from temples. It was observed that the microbial consortium enhanced the digestion of the waste, and it was found to be of high quality without posing any harm to the environment. Sailaja *et al.* reported that the growth rate of plants grown in vermicompost made from temple waste was higher than that of the respective control. In their study, they examined the nutrient status and microbiological activity of the vermicompost prepared. They reported that vermicomposting from flowers contains enzymes that stimulate plant growth and discourage plant pathogens, resulting in a good plant yield. Hence, awareness should be raised among temple authorities, pilgrims, and waste handling personnel to adopt vermicomposting on a large scale for a cleaner environment and financial independence.

4.2 Plant-derived pesticides production

Plant-derived pesticides are a type of bio-pesticides. These pesticides occur in nature and are derived from microbes or plants, which cover a wide spectrum of potential products. Most of the losses in agriculture are related to pests and crop diseases. Thousands of tons of chemical pesticides are being used to eradicate pests, but the excessive use of synthetic pesticides over the years has led to several issues, such as pest resistance and contamination of important natural resources, including water, air, and soil (Kandpal, 2014; Rani *et al.*, 2021). Therefore, to improve the efficiency of crop production and mitigate the food crisis in a sustainable manner while preserving consumers' health, plant-derived pesticides may be a green alternative to synthetic ones (Ghoderao, 2020; Souto *et al.*, 2021). They are economical, biodegradable, ecofriendly, and less hazardous to human health and the environment. Augusto Lopes Souto *et al.* reported a comprehensive study of natural plant products with bioactivity toward insects, which included several classes of molecules, such as terpenes, flavonoids, alkaloids, polyphenols, cyanogenic glucosides, quinones, amides, aldehydes, thiophenes, amino acids, saccharides, and

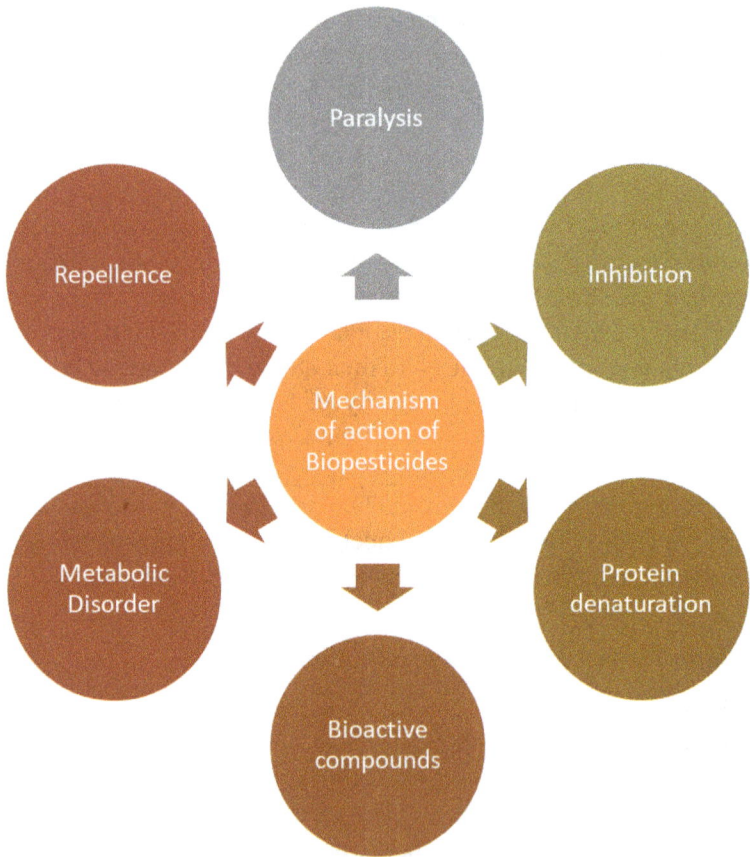

Figure 6.   Mechanism of action of biopesticides.

polyketides (which is not an exhaustive list of insecticidal substances) (Souto *et al.*, 2021). They found compounds for performing important ecological activities in nature, such as antifeedants, attractants, nematicides, fungicides, repellents, insecticides, insect growth regulators, and allelopathic agents, acting as a promising source for novel pest control agents or biopesticides. They described the corresponding toxicology and mechanisms of action for several compounds that act as pesticides. Biopesticides act on pests using several mechanisms of action in a more specific way (Ayilara *et al.*, 2023) (Figure 6).

They also compiled plant-derived metabolites, along with their corresponding toxicology and mechanisms of action, approaches, as well as different strategies to meet the required commercial standards through more efficient methods. They mentioned several factors that limit the commercialization of thes biopesticides. Guido Flamini evaluated the antimicrobial activity of plant extracts and investigated the effectiveness of many plant compounds against bacteria, fungi, protozoa, helminths, and arthropods (Flamini, 2006; Marin & Bautista, 2020). They particularly studied acricides of natural origin and pest (class Arachnida, subclass Acaridi) control in agriculture, veterinary medicine, and human medicine using natural methods (Souto *et al.*, 2021). Golafarin Ghoreishi *et al.* utilized green waste as a substrate to produce biostimulant and biopesticide products through solid-state fermentation (Ghoreishi *et al.*, 2023).

5. Conclusions

The Indian agricultural system has inherently practiced the utilization of AFW since ancient times. Organic farming used to be quite popular until the arrival of the Green Revolution throughout the South Asian region. Later, modern methods of agriculture began attracting the major farmer communities due to faster crop growth and higher productivity. Moreover, since the 1950s, excessive use of chemical compounds such as organophosphate, carbamate, and various organochlorines has caused major damage to the ecosystem by altering soil quality. The discharge of harmful chemicals as various pollutants into the environment reduced soil fertility to a larger extent. Consequently, this affected human health, biodiversity, and soil health. Now, to recover this major damage, it has become essential to reevaluate some of the conventional methods and modernize them in such a way that they can be easily accessible in today's fast-paced world. All types of AFW, as mentioned in this chapter, contain nutrient-rich residues, such as crop remnants, leaves stalks, floral wastes from religious places, APMC waste, domestic waste, food wastes from food processing industries such as pulps and peels of fruits, husks and spent grain of cereals, dregs of coffee, bagasse from sugarcane or sweet sorghum milling, lint, peel, and molasses. If such AFW are properly collected, managed, and utilized for converting them into value-added agricultural products, such as fertilizers, biopesticides, biopolymers, and bio stimulants, then these agricultural practices can play a major role in improving the economy of entre farming communities as well as the nation.

References

Abeliotis, K., Barla, S. A., Detsis, V., & Malindretos, G. (2015). Life cycle assessment of carnation production in Greece. *Journal of Cleaner Production*, *112*, 32–38.

Agapkin, A. M., Makhotina, I. A., Ibragimova, N. A., Goryunova, O. B., Izembayeva, A. K., & Kalachev, S. L. (2022). *The Problem of Agricultural Waste and Ways to Solve it*, p. 022009, IOP Publishing.

Anand, B., Sudha, S., Keshaw, A., & Harit, J. (2013). Value added products from agrowaste. *Recent Research in Science and Technology*, *5*(2), 7–12.

Aqsha, A., Tijani, M., Moghtaderi, B., & Mahinpey, N. (2017). Catalytic pyrolysis of straw biomasses (wheat, flax, oat and barley) and the comparison of their product yields. *Journal of Analytical and Applied Pyrolysis, 125,* 201–208.

Arici, M., Karasu, S., Baslar, M., Toker, O. S., Sagdic, O., & Karaagacli, M. (2016). Tulip petal as a novel natural food colorant source: Extraction optimization and stability studies. *Industrial Crops and Products*, *91*, 215–222.

Ayilara, M. S., Adeleke, B. S., Akinola, S. A., Fayose, C. A., Adeyemi, U. T., Gbadegesin, L. A., Omole, R. K., Johnson, R. M., Uthman, Q. O., & Babalola, O. O. (2023). Biopesticides as a promising alternative to synthetic pesticides: A case for microbial pesticides, phytopesticides, and nanobiopesticides. *Frontiers in Microbiology*, *14*, 1040901.

Bala, S., Garg, D., Inbaraj, B., Singh, R., Kamma, S., Tripathi, M., & Sharma, M. (2023a). Transformation of agro-waste into value-added bioproducts and bioactive compounds: Micro/nano formulations and application in the agri-food-pharma sector. *Bioengineering*, *10*, 152.

Bala, S., Garg, D., Sridhar, K., Inbaraj, B. S., Singh, R., Kamma, S., Tripathi, M., & Sharma, M. (2023b). Transformation of Agro-waste into value-added bioproducts and bioactive compounds: Micro/nano formulations and application in the agri-food-pharma sector. *Bioengineering, 10*(2), 152.

Bierman, P. M. & Rosen, C. J. (1994). Phosphate and trace metal availability from sewage-sludge incinerator ash. *Journal of Environmental Quality, 23*, 822–830.

Devi, S., Gupta, C., Jat, S., & Parmar, M. (2017). Crop residue recycling for economic and environmental sustainability: The case of India. *Open Agriculture, 2*(1), 486–494.

Ding, Y., Zhao, J., Liu, J.W., Zhou, J., Cheng, L., Zhao, J., Shao, Z., Iris, C., Pan, B., Li, X., & Hu, Z. 2021. A review of China's municipal solid waste (MSW) and comparison with international regions: Management and technologies in treatment and resource utilization. *Journal of Cleaner Production, 293*, 126–144.

Enders, M. & Spiegel, M. (1999). Mineralogical and microchemical study of high-temperature reactions in fly-ash scale from a waste incineration plant. *European Journal of Mineralogy, 11*, 763–774.

Fine, P. & Mingelgrin, U. (1996). Release of phosphorus from wasteactivated sludge. *Soil Science Society of America Journal, 60*, 505–511.

Flamini, G. (2006). Acaricides of natural origin. Part 2 — Review of the Literature (2002–2006). *Natural Product Communications, 1*, 1151–1158.

Frossard, E., Sinaj, S., Zhang, L., & Morel, J. (1996). The fate of sludge phosphorus in soil-plant systems. *Soil Science Society of America Journal, 60*(4), 1248–1253.

Fuller, W. H. & Warrick, A. W. (1985). *Soils in Waste Treatment and Utilisation. 1, Land Treatment,* CRC Press Inc., pp. 268.

Ganesh, K. S., Sridhar, A., & Vishali, S. (2022). Utilization of fruit and vegetable waste to produce value-added products: Conventional utilization and emerging opportunities-A review. *Chemosphere, 287*, 132221.

Ghoderao, V. S. (2020). Effect of biopesticide from custard apple seeds on white mealy bugs. *International Research Journal of Engineering and Technology, 7*(5), 5265–5267.

Ghoreishi, G., Barrena, R., & Font, X. (2023). Using green waste as substrate to produce biostimulant and biopesticide products through solid-state fermentation. *Waste Management, 159*, 84–92.

Indiastat. 2020. Area and Production of Total Flowers in India (1993–1994 to 2019–2020 3rd advance estimates).

Jakobsen, P. & Willett, I. (1986). Comparisons of the fertilizing and liming properties of lime treated sewage sludge with its incinerated ash. *Fertil Research, 9*, 187–197.

Kamusoko, R., Jingura, R. M., Parawira, W., & Chikwambi, Z. (2021). Strategies for valorization of crop residues into biofuels and other value-added products. *Biofuels, Bioproducts and Biorefining, 15*(6), 1950–1964.

Kandpal, V. (2014). Biopesticides. *International Journal of Environmental and Rural Development*, *4*(2), 191–196.

Kaza, S., Yao, L., Bhada-Tata, P., & Van Woerden, F. (2018). *What a Waste 2.0: A Global Snapshot of Solid Waste Management to 2050*. The World Bank Publication, Washington DC 20433, 292.

Khedkar, R. & Singh, K. (2017). Food industry waste: A panacea or pollution hazard? *Paradigms in Pollution Prevention*, 35–47. https://doi.org/ 10.1007/978-3-319-58415-7_3.

Khedkar, R. & Singh, K. (2018). Food industry waste: *A panacea* or pollution hazard? In Jindal, T. (ed.) *Paradigms in Pollution Prevention*, pp. 35–47.

Lohan, S. K., Jat, H. S., Yadav, A. K., Sidhu, H. S., Jat, M. L., Choudhary, M., & Sharma, P. C. (2018). Burning issues of paddy residue management in northwest states of India. *Renewable and Sustainable Energy Reviews 81*, 693–706.

Mahindrakar, A. (2018) Floral waste utilization- a review, *International Journal of Pure and Applied Biosciences, 6*(2), 325–329.

Mane, T. T. & Raskar, S. S. (2012). Management of agriculture waste from market yard through vermicomposting. *Research Journal of Recent Sciences.*, 1, 289–296.

Marin, M. B. & Bautista, C. V. (2020). Weed extracts as potential biopesticides against Cabbage black rot in an upland of Southern Philippines. *IOP Conference Series: Earth and Environmental Science*, *449*(1), 012027.

Masure, P. S. & Patil, B. M. (2014). Extraction of waste flowers. *International Journal of Engineering Research and Technology*, *3*, 43–44.

Mellbye M., Hemphill D., & Volk V. (1982). Sweet corn growth on incinerated sewage, sludge-amended soil. *Journal of Environmental Quality*, *11*, 160–163.

Mishra, R. & Mohanty, K. (2020). Kinetic analysis and pyrolysis behaviour of waste biomass towards its bioenergy potential. *Bioresource Technology*, *311*, 123480.

Mlonka-Mędrala, A., Evangelopoulos, P., Sieradzka, M., Zajemska, M., & Magdziarz, A. (2021). Pyrolysis of agricultural waste biomass towards production of gas fuel and high-quality char: Experimental and numerical investigations, *Fuel, 296*, 120611.

NWOSIBE Patrick O, E. P. N. & J, N. T. U. A. (2020). Production of organic fertilizer using biodegradable waste. *International Journal of Scientific & Engineering Research*, *11*(9), 431–438.

Pryshliak, N. & Tokarchuk, D. (2020). Socio-economic and environmental benefits of biofuel production development from agricultural waste in Ukraine. *Environmental Socio-economical Studies*, *8*(1), 18–27.

Rai, J. P. N. & Saraswat, S. (2023). *Green Technologies for Waste Management: A Wealth from Waste Approach* (1st ed.). CRC Press., CRC Press Taylor & Francis Group.

Rani, L., Thapa, K., Kanojia, N., Sharma, N., Singh, S., Grewal, A. S., Srivastav, A. L., & Kaushal, J. (2021). An extensive review on the consequences of chemical pesticides on human health and environment. *Journal of Cleaner Production, 283,* 124657.

Rehman, S. U., De Castro, F., Aprile, A., Benedetti, M., & Fanizzi, F. P. (2023). Vermicompost: Enhancing plant growth and combating abiotic and biotic stress. *Agronomy, 13*(4), 1134.

Rudra, S. G., Nishad, J., Jakhar, N., & Kaur, C. (2015). Food industry waste: Mine of nutraceuticals, IInt. *Journal of Environmental Science and Technology, 4*(1), 205–229.

Sharma, R. K. (2021). Floral waste management & opportunities. *Just Agriculture, 113.*

Sharmila Rani, P. K. S. & Menka, B. (2020). Technology for utilization of floral waste and corresponding products — A review. *International Journal of Scientific & Engineering Research, 11*(9), 1555–1558.

Shu, H., Zhang, P., Chang, C.-C., Wang, R., & Zhang, S. (2015). Agricultural waste. *Water Environment Research, 87*(10), 1256–1285.

Singh, A. & Dubey G. (2022). Agricultural waste management: Problems and treatments. *JIRMF 8*(4), 148–152.

Sommers, L. (1977). Chemical composition of sewage sludges and analysis of their potential use as fertilizers. *Journal of Environmental Quality, 6,* 225–232.

Soundarya, S., Radhika, P., Desireddy, S., & Supriya, K. (2021). An analysis of pattern of floral waste generated and disposal in Hyderabad City of Telangana state. *Asian Journal of Agricultural Extension, Economics & Sociology, 39*(11), 414–420.

Souto, A. L., Sylvestre, M., Tölke, E. D., Tavares, J. F., Barbosa-Filho, J. M., & Cebrián-Torrejón, G. (2021). Plant-derived pesticides as an alternative to pest management and sustainable agricultural production: Prospects, applications and challenges. *Molecules, 26*(16), 4835.

Tchonkouang, R. D., Onyeaka, H., & Miri, T. (2023). From waste to plate: Exploring the impact of food waste valorisation on achieving zero hunger. *Sustainability, 15*(13), 10571.

Tesfai, M. & Dresher, S. (2009). Assessment of benefits and risks of landfill materials for agriculture in Eritrea. *Waste Management, 29*(2), 851–858.

Vigneswaran, C., Pavithra, V., Gayathri, V., & Mythili, K. (2015). Banana fiber: Scope and value added product development. *Journal of Textile and Apparel, Technology and Management, 9*(2), 2001–2008.

Waghmode, M. S., Gunjal, A. B., Nawani, N. N., & Patil, N. N. (2018). Management of floral waste by conversion to value-added products and their other applications. *Waste and Biomass Valorization, 9*(1), 33–43.

Williams, M. Pedersen, B., & Jorgensen S. (1976). Composition and reactivity of ash from sewage sludge. *AMBIO Journal of Human Environment, 5,* 32–35.

Zhang, F., Yamasaki, S., & Kimura, K. (2001). Rare earth element content in various waste ashes and the potential risk to Japanese soils. *Environment International, 27,* 393–398.

Zhu, Y., Luan, Y., Zhao, Y., Liu, J., Duan, Z., & Ruan, R. (2023). Current technologies and uses for fruit and vegetable wastes in a sustainable system: A review. *Foods, 12*(10), 1949.

Chapter 8

Clean Technologies for Minimizing Waste from Manufacture of Dairy Products

Edna Regina Amante[*], Lorena Samara Gama Pantoja,
Antonio Manoel da Cruz Rodrigues, and
Luiza Helena Meller da Silva

*Programa de Pós-Graduação em Ciência e Tecnologia de Alimentos,
Universidade Federal do Pará, Rua Augusto Correa S/N,
Guamá, Belém, Pará, 66075-900, Brazil*
eamante.1957@gmail.com

Abstract

The nutritional value of milk and dairy products is manifested in the nature of effluents with a high organic load, resulting from the water used in equipment cleaning processes, as well as in the by-products of the production of cheese, butter, and other derivatives. On the other hand, the dairy industry also presents itself on different scales, ranging from family-owned businesses to large conglomerates; in all cases, this sector coexists with the need to reduce the pollutant load of its production process. In general, the industry complies with environmental regulations through the waste treatment system. In this chapter, from the perspective of clean technologies, systems for minimizing and valuing waste are suggested for the main industrial waste in the dairy sector through *an a priori* analysis based on a comparison between the investment required in each proposal and the environmental and economic benefits. The proposals are built based on published scientific work on the possibilities of recovering this waste. The suggestions from the different systems are offered as a contribution for this productive sector's vision of transforming what is considered polluting into a new income opportunity.

Keywords: Dairy industry, wastes minimization, wastes valorization.

1. Introduction

The efficiency of result multiplication from a scientific work can be evaluated by its applications for the benefit of society, in an anthropo-geocentric view, to humans and the environment. The food industry, which is widely spread across the planet and follows specific regional cultures and traditions, requires effective measures for sustainable production. While waste treatment systems are concerned with meeting the parameters required by environmental regulatory agencies, aiming to reach the required levels, the valorization of raw materials can both reduce the volumes of waste and effluents treated in conventional systems and contribute to the generation of new products and income, both from soluble and suspended solids in effluents and from solid waste from different sources in each production sector.

The justifications for food production chains being considered the cause of negative impact on the environment are based on the traditional consideration that in order to produce, products and waste are generated and that waste must be treated, aiming to minimize the environmental impact. However, sustainability in all productive sectors, including the food industry, requires a rethink based on the large volume of current knowledge and advanced technologies available.

Existing technologies need to be reviewed by the production sector and transferred efficiently. These technology transfers are possible on any industrial scale, as long as there is compatibility between the proposed solutions and the technological reality of producing companies and communities. Seeking to achieve this compatibility, projects that value raw materials and minimize waste must be specific and carried out by the particular type of industry in a given production sector, involving a range of possibilities for the complete valorization of raw materials.

The dairy industry is an important example of this diversity, where proposals have been presented and where decisions about valuing raw materials and minimizing waste can be a reality as long as there is an organization of the sector and environmental actors, increasing the permeability of information and initiatives to consider the problem in a sectorized way, regardless of the production scale.

2. The Dairy Industry as an Example of Applying Systems Models for Sustainable Production

The bovine milk and derivatives sector can represent an important example, where whey, which was considered a polluting waste until less than three decades ago, has come to represent a by-product with high-added-value derivatives and thus has contributed to reducing environmental risks in cheese production (Farkas *et al.*, 2019; Pantoja *et al.*, 2022). On the other hand, this is also a typical example of the effects of the diversity of scales in this sector, with large companies coexisting with small rural businesses owning only a few heads of cattle, where whey has not yet found applications and, even today, still represents a risk to the environment. However, this scenario presents a great opportunity for implementing changes, as the options for using whey for small producers have been studied (Cunha *et al.*, 2008) and should be encouraged since small properties are common in countries that have low per capita income and malnutrition.

Proposals for the models of waste minimization in this sector must therefore cover suggestions applicable to different scales because, according to the OCED-FAO Outlook 2023–2032, there is a great diversity of scales in all regions. In low- and lower-middle-income countries, most of the production is consumed in the form of fresh dairy products. The consumption of fresh dairy products per capita is expected to be high in India and Pakistan but low in China. In Europe and North America, the overall per capita demand for fresh dairy products is stable or declining, but the composition of demand has been shifting over the past few years toward dairy fats, such as full-fat drinking milk and cream.

The share of processed dairy products, especially cheese, in overall consumption of milk solids is expected to be closely related to different income ranges, with variations due to local preferences, dietary constraints, and urbanization. The largest share of total cheese consumption, the second-most consumed dairy product, occurs in Europe and North America, where per capita consumption is expected to continue to increase over the projection period. Consumption of cheese will also increase in regions where it has not traditionally been part of the regional diet. In Southeast Asian countries, an increase in urbanization and income has resulted in more "away-from-home" eating, including fast foods such as burgers and pizzas (OECD-FAO, 2023–2032).

The projection for cheese consumption in 2032 ranges from less than 2 kg/person least developed countries and East Asia to 20 kg/person in North America (OECD-FAO, 2023–2032). These data can be used for projecting the production of whey, considering that in the European Union, dairy industries produce, on average, 2.5 L of wastewater per liter of processed milk as well as about 9–10 L of cheese whey per kg of cheese (Berlese *et al.*, 2019; *Eurostat*, 2018). This does not constitute an environmental problem where the industrialization of whey has been consolidated. Therefore, a differentiated approach needs to be adopted in low- and middle-income countries where derivatives are produced, especially cheese.

Seeking to contribute to the minimization of environmental impact in the dairy sector, it is possible to evaluate the factors that interfere with consumers' growing justifications for replacing milk with plant-based drinks, alleging environmental issues for ceasing consumption of milk and dairy products, in addition to habits that are considered healthier and restrictions due to lactose intolerance and allergies. In this aspect, environmental legislation could have a strong impact on the future development of dairy production. GHG emissions from dairy activities make up a high share of total emissions in some countries (e.g., New Zealand and Ireland), and more stringent environmental policies and initiatives, such as the Pathways to Dairy Net Zero launched in September 2021 by the dairy industry, could affect the level and nature of dairy production to curb such emissions.

Glavas and Fitzgerald (2020) presented a historic example of how the American dairy industry has achieved exemplary environmental results by integrating all segments of the production chain, from farms to distributors and processing units, thus showing that joint commitment contributes to success.

The basic steps followed in milk processing, adapted from Pantoja *et al.* (2022), are shown in Figure 1, where some suggestions for the valorization of by-products in the production chain are included.

Wastewater generation in each stage of the production process is not included in the synthesis shown in Figure 1. However, if we consider the

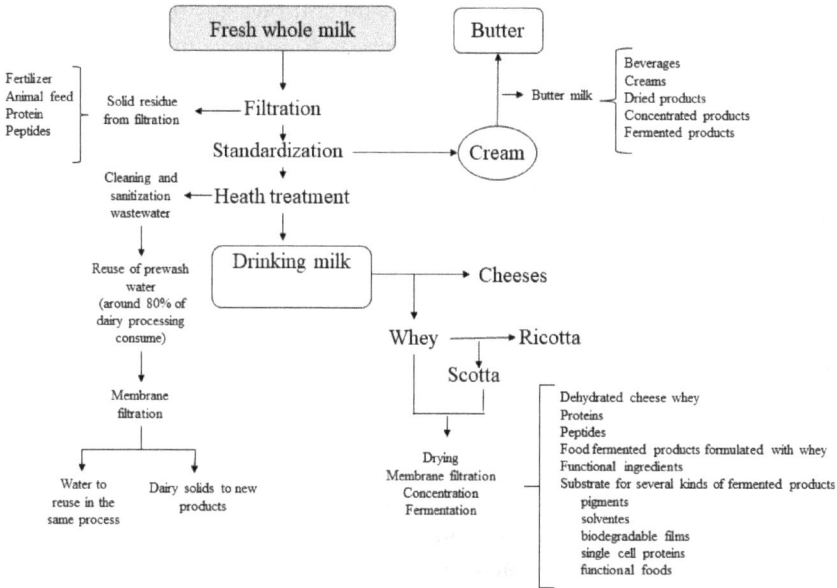

Figure 1. Proposals for the minimization of residues in the preparation of the main products of the milk production chain (adapted from Pantoja *et al.*, 2022).

example of the European Union, where 2.5 L of wastewater is generated per liter of milk processed and that the composition of that wastewater can be valorized, a different scenario emerges regarding clean technology concepts, where each step of milk processing should be investigated to minimize waste production as well as water and energy consumption.

Knowledge about the different stages of the processes, both at the farm and industry levels, as well as the integration of actors, as clearly explained in the work by Glavas and Fitzgerald (2020), can contribute not only to the minimization of environmental impacts but also to the increase in income and social benefits in the sector under study.

What is in the hands of the scientific community in the food sector and that can contribute to suggestions for improvements in milk processing is the evaluation of the chemical composition of liquid and solid wastes as well as the performance at the different stages. In the case of the dairy industry, the use of detergents and sanitizers that can be purified and reused, in addition to using different pipes and tubes, can significantly change the pollutant load.

Meanwhile, knowledge of effluents and solid wastes that do not contain sanitizers, detergents, or other materials foreign to milk can lead to the development of new products, as shown in Figure 1.

In the milk processing industry, waste treatment aims to reduce the pollutant load by reaching the parameters required by local environmental legislation. However, in the treatment systems adopted, effluents are not valorized, which differs from actions that emphasize clean technologies, where the composition and properties of liquid and solid waste are taken into consideration and thus justify the suggested applications (Figure 1). Taking these initiatives into account, proposals for application are presented for each type of waste in a simplified system, considering the level of investment and the environmental and economic benefits (Oliveira *et al.*, 2013; Rezzadori *et al.*, 2012).

Regarding decision-making, the distances between the production units, the difference in scales, the technological realities of the original waste-generating enterprise, the initiative to form groups to reach the desired scales according to each suggestion for application, and the characteristics of the waste-generating unit will all contribute to the decision on the geographic locations where the by-products will be processed.

In addition to the two above-mentioned works and others published, one great example of this diversity draws attention to the dairy industry, where producers may coexist with only a few animal units in the cattle herd (where processing occurs on smaller rural properties) or with the manufacture of dairy products in large cooperatives (processing millions of liters of bovine milk each day), both of which possibly need environmental solutions. To exemplify the application of the proposed model, this diversity of the dairy industry will be used to present possible solutions for reducing the environmental impact of agricultural and agro-industrial sectors.

Taking into account the suggested possibilities of application, although many production processes are traditional, measures are taken to ensure that new environmentally friendly technologies are adopted, since in many processes, the classic steps can represent more environmental damage as they use processing conditions that are incompatible with the objectives of clean technologies.

An approach to the main technologies adopted and the possibilities of destination for each type of waste will be first presented as systems. Subsequently, the systems will be listed and compared in terms of investment range and economic and environmental benefits. System 1, in all suggestions, will always correspond to the lowest level of valorization and only meet what is required by environmental regulatory agencies. The systems will be listed until suggestions for the use of each liquid or solid waste are exhausted.

Suggested systems for waste minimization and recovery can be presented according to the required investment and their environmental and economic benefits. The economic benefit is related to the main products, where "low" (L) = x, means the same economic return as that of commercial milk, "medium" (M) $\leq 5x$ that of commercial milk, and "high" (H) $\geq 5x$ that of commercial milk. As for the environmental benefit: low (L): values of the parameters of environmental evaluation are reduced by 80%; medium (M): values of the parameters of environmental evaluation are reduced in 90%; high (H): values of the parameters of environmental evaluation are reduced in more than 90%. Similarly, in the case of economic investment: low (L): investment in the installation of the system is the same as that for the conventional waste treatment; medium (M): investment in the installation of the system is at least twice higher than that for the conventional waste treatment; high (H): the investment with the proposed system is equal to or greater than the industrial unit generating the waste (Oliveira *et al.*, 2013; Rezzadori *et al.*, 2012).

Table 1 shows the principal operations, economic benefit, environmental benefit, and economic investment of the proposed systems for the valorization of solid waste from fresh whole-milk filtration. Five systems are suggested, where the first system always considers traditional treatment practices or the destination of the waste under evaluation. This solid residue is rarely considered for recovery, but it constitutes protein material despite the low percentages in the volume of the processed milk. System 1 considers disposal in a landfill. System 2 considers composting or use as a source of nitrogen in the preparation of fertilizer, which consists of an application with low economic investment, high environmental benefit, and medium economic benefit, already surpassing system 1, despite its simplicity. Through the use of a process that guarantees safe consumption, system 3 suggests application in animal feed, with a high environmental return, medium economic benefit, and medium investment.

Systems 4 and 5, on the other hand, require high economic investment, resulting in high economic benefit and medium environmental benefit, since effluents are generated in the process. Always drawing attention to the fact that these effluents must be characterized, which

Table 1. Principal operations, economic benefit, environmental benefit, and economic investment of proposed systems for valorization of solid waste from fresh whole milk filtration.

Systems*	Original solid residue	Final solid residue	Final liquid residue	Transportation	Milling	Centrifugation	Drying	Hydrolysis	Heat treatment	Membrane filtration	Composting	Concentration	Economic benefit	Environmental benefit	Economic investment
									Operations					**Environ-**	
1	x			x									L	L	L
2	x			x							x		M	H	L
3	x			x	x	x	x		x				M	H	M
4	x	x	x	x	x	x	x	x	x	x		x	H	M	H
5	x	x	x	x	x	x	x		x	x		x	H	M	H

Notes: *1 – Landfill disposition; 2 – organic fertilizer; 3 – animal feed; 4 – peptides; 5 – protein. H – High; L – Low; M – Median.

could result in new products, it is thus important to exhaust all return possibilities in each of the suggested systems. As for the scale to enable the use of each system, a second step lies in knowing the volumes produced by each processing unit and the production location, making cooperation between processing units possible.

Table 2 shows the principal operations, economic benefit, environmental benefit, and economic investment of the proposed systems for the valorization of buttermilk, an important effluent in the dairy industry. Five systems are presented, where system 1 constitutes traditional effluent treatment systems that comply with environmental legislation but do not value soluble and suspended solids. The investment is low since, in an effluent treatment unit, there is no economic return, and the environmental return is that it only complies with environmental legislation. We should always consider that this effluent has a high organic load and that, in some legislation, a certain percentage of reduction is sufficient to comply with the regulation. System 2 suggests the use of this nutritive effluent for beverage production, which can be treated as a by-product. In this case, high investment is required, while achieving a medium environmental benefit and high economic benefit, with the detail that this process can be suitable for a range of scales, from small to large companies. This effluent can also be used to prepare cream (system 3) by using tangential filtration and formulating a cream from the retentate. The economic investment is high, the economic benefit is high, and the environmental benefit is medium.

System 4 proposes the application to protein production using tangential filtration, concentration, and drying. High investment is required, but it results in a product with high economic and environmental benefits. However, it also depends on research and attempts to value all the separate fractions for new products with different purposes. From the protein produced in system 5, there may be an option for the preparation of peptides; in this case, equally, the investment is high, but the environmental and economic benefits are high. Both system 4 and system 5 have a profile for large-scale application; however, the possibility of a consortium between smaller companies producing this effluent is not ruled out, bringing broad benefits, which will depend on the organization of the sector or the producing region.

Table 3 shows the principal operations, economic benefit, environmental benefit, and economic investment of the proposed systems for the valorization of cleaning and sanitization of wastewater. A large part of the water consumption in the dairy sector is intended for cleaning and

Table 2.   Principal operations, economic benefit, environmental benefit, and economic investment of proposed systems for valorization of butter milk.

Systems*	Original liquid residue	Final liquid residue	Transportation	Membrane filtration	Concentration	Filtration	Mixture	Heat treatment	Hydrolysis	Drying	Wastewater treatment	Economic benefit	Environmental benefit	Economic investment
												Operations		
1	x		x								x	L	L	L
2	x	x	x			x	x	x				H	M	H
3	x	x	x	x			x				x	H	M	H
4	x	x	x	x	x					x	x	H	H	H
5	x	x	x	x	x				x		x	H	H	H

Notes: *1 – Conventional wastewater treatment; 2 – beverages; 3 – creams; 4 – protein; 5 – peptides H – High; L – Low; M – Median.

Table 3. Principal operations, economic benefit, environmental benefit, and economic investment of proposed systems for valorization of cleaning and sanitization wastewater.

Systems*	Original liquid residue	Transportation	Membrane filtration	Concentration	Hydrolysis	Heat treatment	Drying	Wastewater treatment	Economic benefit	Environmental benefit	Economic investment
						Operations					
1	x	x						x	L	L	L
2	x	x	x					x	L	M	M
3	x	x	x	x		x	x		M	H	H
4	x	x	x	x				x	H	M	H
5	x	x		x	x		x	x	H	M	H

Notes: *1 – Wastewater treatment system; 2 – water for industrial reuse; 3 – solids from wastewater filtration — valorization for new products — animal feed; 4 – proteins; 5 – peptides. H – High; L – Low; M – Median.

sanitizing operations, which requires actions to ensure that these waters can be recirculated in the process itself as well as when the soluble and suspended solids are made up of milk or milk residues. It also involves the preparation of derivatives, not mixed with sanitizers and detergents, to provide a safe destination for these materials.

In system 1, these effluents are treated; in system 2, it is already considered that the effluent emission lines are separate and that sanitizers and detergents do not follow the same flow as the effluents that contain the pre-wash waters, which correspond to more than 90% of the volume of water used to clean equipment in this sector; thus, water can be filtered and solids retained for new applications, which is proposed in systems 3, 4, and 5. The systems propose diverse applications, from simple ones such as an ingredient for animal feed to the possibility of producing proteins and peptides with a broad spectrum of use, which means a significant difference between treating effluents and degrading these compounds or deciding to separate them.

Table 4 shows the principal operations, economic benefit, environmental benefit, and economic investment of the proposed systems for the valorization of whey and ricotta whey. There are a large number of studies (Rivas *et al.*, 2010; Pasotti *et al.*, 2017; Pandey *et al.*, 2019; Papademas & Kotsaki, 2019; Shivanna & Nataraj, 2020) dedicated to the valorization of whey, which allows us to suggest at least 10 systems based on these works, as well as the availability of whey derivatives already commercialized. Despite this development, it is still common to discard whey in some regions, especially in low-income countries (Rivas *et al.*, 2010; Pathak *et al.*, 2016). Therefore, although system 1 has been suggested, producers can still be found offering liquid whey for animal feed or even discarding it irregularly into the environment. With the value reached by whey, this is not the prevailing reality, where it can be said that whey is one of the greatest examples to justify that all liquid and solid waste must be recognized for its proper valorization, as whey has already been considered an environmental villain, but today, its derivatives are worth more than traditional dairy products.

Thus, systems 2–10 suggested for whey already represent reality, and this fact is directly related to investments in research and innovation, where full use of other residues from the dairy industry has a highly positive future perspective. All the suggestions offered (Tables 1–4) need to be plotted according to the scale of each production unit, with an assessment of the real benefits of using each application suggested.

Table 4. Principal operations, economic benefit, environmental benefit, and economic investment of proposed systems for valorization of whey and ricotta whey.

	Original liquid residue	Final liquid residue	Transportation	Fermentation	Centrifugation	Concentration	Distillation	Precipitation	Hydrolysis	Heat treatment	Membrane filtration	Wastewater treatment	Drying	Economic benefit	Environmental benefit	Economic investment
Systems*																
1	x	x	x											L	L	L
2	x	x	x			x					x		x	H	H	H
3	x	x	x		x	x		x			x	x	x	H	M	H
4	x	x	x		x	x			x	x	x	x	x	H	M	H
5	x	x	x	x								x		H	H	M
6	x	x	x	x	x	x						x	x	H	H	M
7	x	x	x	x	x		x					x		H	M	H
8	x		x			x					x		x	M	H	H
9	x	x	x	x	x							x	x	H	M	H
10	x	x	x			x		x			x	x	x	H	M	H

Notes: *1 – Wastewater treatment; 2 – whey powder; 3 – whey protein; 4 – peptides; 5 – fermented foods; 6 – pigments using whey as substrate; 7 – solvents by whey fermentation; 8 – biodegradable films; 9 – single cell protein; 10 – functional foods. H – High; L – Low; M – Median.

The suggestions offered in this work certainly do not exhaust all possible uses; however, they are intended to serve as an exercise for the production sector to reflect on the value of what is being discarded as waste and also as an idea of the possibility of minimizing environmental impact, thus adding value to what is usually referred to as "waste." It is possible to note that specific knowledge of the sector, as well as of the processing steps in the preparation of each product, is essential for the success of these actions in clean technologies.

Each production unit can build a similar spectrum of possibilities and practice the habit of mapping water consumption, generation of effluents and solid waste, and providing production spreadsheets, as is commonly

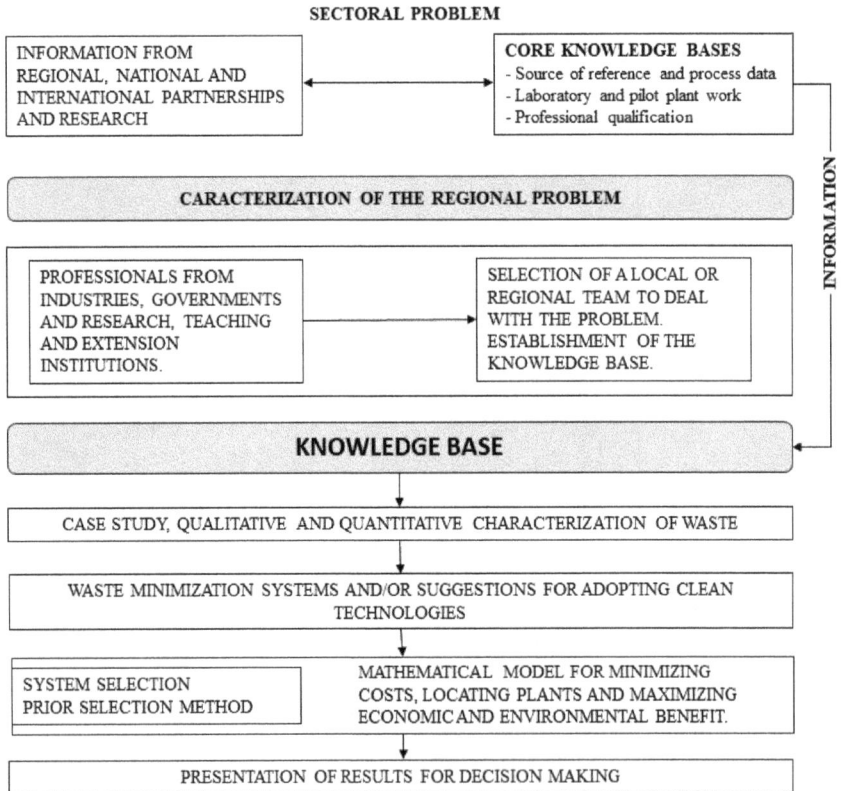

SECTORAL PROBLEM

INFORMATION FROM REGIONAL, NATIONAL AND INTERNATIONAL PARTNERSHIPS AND RESEARCH	CORE KNOWLEDGE BASES - Source of reference and process data - Laboratory and pilot plant work - Professional qualification

CARACTERIZATION OF THE REGIONAL PROBLEM

PROFESSIONALS FROM INDUSTRIES, GOVERNMENTS AND RESEARCH, TEACHING AND EXTENSION INSTITUTIONS.	SELECTION OF A LOCAL OR REGIONAL TEAM TO DEAL WITH THE PROBLEM. ESTABLISHMENT OF THE KNOWLEDGE BASE.

KNOWLEDGE BASE

CASE STUDY, QUALITATIVE AND QUANTITATIVE CHARACTERIZATION OF WASTE

WASTE MINIMIZATION SYSTEMS AND/OR SUGGESTIONS FOR ADOPTING CLEAN TECHNOLOGIES

SYSTEM SELECTION PRIOR SELECTION METHOD	MATHEMATICAL MODEL FOR MINIMIZING COSTS, LOCATING PLANTS AND MAXIMIZING ECONOMIC AND ENVIRONMENTAL BENEFIT.

PRESENTATION OF RESULTS FOR DECISION MAKING

INFORMATION

Figure 2. Proposal for the formation of knowledge bases for solving sector-specific environmental problems.

carried out with the registration of the main products. From this quantification, it is possible to make the decision to use any of the systems proposed, having an idea of how many new products it is possible to create from a specific waste. It turns out that the scale size is a key factor in this decision, as some proposals are not possible on small scales, where, depending on the concentration of small processing units, a group of companies interested in waste valorization can be formed. In this case, simple modeling can decide the best processing location, depending on the quantities produced and transported, as well as the scale of the new unit, and thus make the processing more viable than when using only one processing unit.

A sectoral organization can contribute as a source of information for all sectors, not just the one exemplified in Figure 2. With organized and supported sectors, the efficiency of waste minimization can be optimized.

A combination of graphs of production scale per waste generated can be constructed for each proposed system, where decisions can be made, including compliance with environmental legislation, income generation, and possibilities for creating new products.

3. Conclusion

The establishment of simplified and accessible methods for the production sector can help all actors involved contribute to minimizing environmental impact, as shown in the example presented for the dairy sector.

While waste treatment seeks to meet the parameters established by environmental legislation, the incorporation of clean technology concepts in the production sector offers an important alternative, both for reducing environmental impact and for generating new products and income.

The analysis of the different possibilities for waste recovery, according to the technological, economic, and regional realities of each company, or a set of them, expands the possibilities in decision-making between traditional waste treatment and new business opportunities, reducing the environmental impact.

The construction of the set of suggestions can be expanded using the accumulated knowledge about each waste generated. This is valid and should be encouraged not only in the example presented but for all sectors because, in addition to the simplicity of the analysis, it offers itself as an opportunity for decision-making in the productive sector, unlike the option of merely treating waste, but rather to minimize it, reducing operations in waste treatment units and expanding opportunities in the productive sector.

References

Berlese, M., Corazzin, M., & Bovolenta, S. (2019). Environmental sustainability assessment of buffalo mozzarella cheese production chain: A scenario analysis. *Journal of Cleaner Production, 238*, 117922.

Cunha, T. M. C., Castro, F. P., Barreto, P. L. M., Benedet, H. D., & Prudêncio, E. S. (2008). Physico-chemical, microbiological and rheological evaluation of dairy beverage and fermented milk added of probiotics. *Semina – Ciências Agrárias, 109*, 113–116.

Eurostat (2018). *Agriculture, Forestry and Fishery Statistics*, 2018 Edition.

Food and Agriculture Organization (FAO) (2018). The FAO blue growth initiative: Strategy for the development of fisheries and aquaculture in eastern Africa. Available at: https://www.fao.org/3/i8512en/I8512EN.pdf (Accessed 3 February 2024).

Farkas, C., Rezessy-Szabó, J. M., Gupta, V. K., Bujna, E., Pham, T. M., Pásztor-Huszár, K., Friedrich, L., Bhat, R., Thakur, V. K., & Nguyen, Q. D. (2019). Batch and fed-batch ethanol fermentation of cheese-whey powder with mixed cultures of different yeasts. *Energies, 12*, 4495.

Glavas, A. & Fitzgerald, E. (2020). The process of voluntary radical change for corporate social responsibility: The case of the dairy industry. *Journal of Business Research, 110*, 184–201.

OECD-FAO (2023–2032). Outlook. https://www.oecd-ilibrary.org/docserver/08801ab7-en.pdf?expires=1706810147&id=id&accname=guest&checksum=81F90EBFDF75529583435DB89EEF2701 (Consulted 1 February 2024).

Oliveira, D. A., Benelli, P., & Amante, E. R. (2013). A literature review on adding value to solid residues: Egg shells. *Journal of Cleaner Production, 46*, 42–47.

Pandey, A., Mishra, A. A., Shukla, R. N., Dubey, P. K., & Vasant, R. K. (2019). Development of the Process for Whey Based Pineapple Beverage. *International Journal of Current Microbiology and Applied Sciences, 8*, 3212–3228.

Pantoja, L. S. G., Amante, E. R., Rodrigues, A. M. C., & Silva, L. H. M. (2022). World scenario for the valorization of byproducts of buffalo milk production chain. *Journal of Cleaner Production, 364*, 132605.

Papademas, P. & Kotsaki, P. (2019). Technological utilization of whey towards sustainable exploitation. *Advances in Dairy Research, 7*, 231.

Pasotti, L., Zucca, S., Casanova, M., Micoli, G., De Angelis, M. G. C., & Magni, P. (2017). Fermentation of lactose to ethanol in cheese whey permeate and concentrated permeate by engineered *Escherichia coli*. *BMC Biotechnology*, *17*, 1–12.

Pathak, U., Das, P., Banerjee, P., & Datta, S. (2016). Treatment of wastewater from a dairy industry using rice husk as adsorbent: Treatment efficiency, isotherm, thermodynamics, and kinetics modelling. *Journal of Thermodynamics*, *2016*, 3746316.

Rezzadori, K., Benedeti, S., & Amante, E. R. (2012). Proposals for the residues recovery: Orange waste as raw material for new products. *Food and Bioproducts Processing*, *90*, 606–614.

Rivas, J., Prazeres, A. R., Carvalho, F., & Beltrán, F. (2010). Treatment of cheese whey wastewater: Combined coagulation — flocculation and aerobic bio-degradation. *Journal of Agricultural and Food Chemistry*, *58*, 7871–7877.

Shivanna, S. K. & Nataraj, B. H. (2020). Revisiting therapeutic and toxicological fingerprints of milk-derived bioactive peptides: An overview. *Food Bioscience*, *38*, 100771.

Chapter 9

Recovery of Elements Using Green Plants for Potential Valorization of Waste Metals

Mamatha Bhanu LS[*,‖], Anindita Mitra[†], Aswetha Iyer[‡],
Dharmendra K. Gupta[§], and Soumya Chatterjee[¶,**]

[*]Department of Biotechnology, Yuvaraja's College, University of Mysore,
Manasagangotri, Mysuru-570006, Karnataka, India

[†]Bankura Christian College, Bankura-722101, West Bengal, India

[‡]Department of Biotechnology, Karunya Institute of Technology and
Sciences (Deemed to be University), Coimbatore-641114, India

[§]Ministry of Environment, Forest and Climate Change, Indira
Paryavaran Bhavan, Aliganj, Jorbagh Road, New Delhi-110003, India

[¶]Biodegradation Technology Division, Defence Research Laboratory,
DRDO, Tezpur 784001, Assam, India

[‖]ls.mamatha@gmail.com

[**]drlsoumya@gmail.com

Abstract

Many manufacturing items require specific elements, including, for
example, low-carbon technologies such as wind turbines, electric cars,
and catalytic converters. The growing global population and aspirations
for a better life have raised concerns about the security and accessibility

of valuable elements. These economically important metals need appropriate, environment-friendly recovery technology from increasing waste repositories, which will help both in hazard avoidance and as a potential source for industry. Phytoextraction, also known as phytomining, is a green and innovative method for recovering metals from waste, mainly focusing on resource supply, while contributing to hazard mitigation. In this context, plants have been found to be capable of phytomining platinum group metals (PGMs) to create stable metal nanoparticles that are useful in various industrial reactions. The use of mine tailings or waste mining waters for metal adsorption in plant beds is followed by controlled pyrolysis to produce stabilized PGM nanoparticles for heterogeneous catalysts. The proposed solution to the global issue of metal depletion can lead to the creation of a new range of naturally derived catalysts. This review explores a multidisciplinary approach to evaluating metal sustainability, highlighting key aspects such as metal lifecycle analysis, waste sources, phytoextraction, and potential green chemical applications.

Keywords: Green chemistry, phytoremediation, metal recovery, phytoextraction, toxic metals.

1. Introduction

The contamination of the environment due to industrialization and urbanization has significantly increased the environmental abundance of heavy metals, raising global concerns (Suman *et al.*, 2018; Ashraf *et al.*, 2019). Human activities such as mining ores, emitting gases, using pesticides, and producing municipal waste have significantly contributed to the addition of pollutants to soil, water, and the atmosphere over time (Shah and Daverey, 2020). Mining is the excavation of the original source of all metallic elements, where new industrially important metals must be obtained through the exploration and exploitation of ore bodies. As society progressed and the use of these metals increased with time, the elements became increaingly more dispersed throughout the "technosphere." Recapturing and reusing these metals is not easy with conventional technologies and is often costly. Indiscriminate pollution caused by heavy metals is apparent, as these accumulate in food chains, causing harm to plants, animals, and humans (Nedjimi, 2009). Currently, many elements

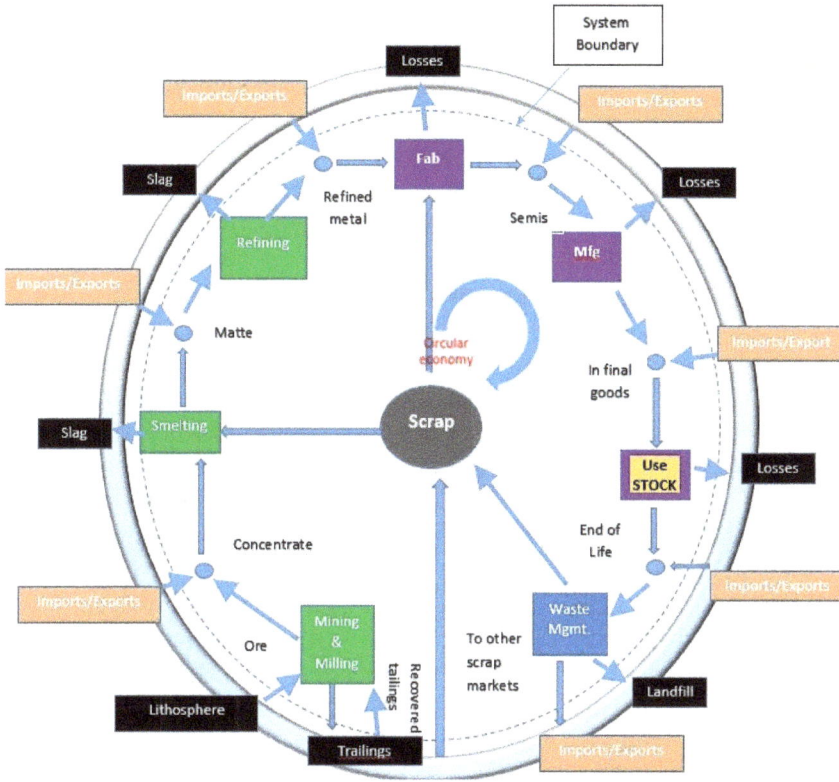

Figure 1. Lifecycle of metals: mining and milling; smelting, refining, fabricating (Fab.), manufacturing (Mfg.), use, and waste management and recycling (Waste Management) (adapted from Hunt *et al.*, 2014).

have low recycling rates, with the efficiency of recycling being primarily determined by the collection of waste metal immediately after use. For example, platinum (Pt) and lead (Pb) have established recycling routes due to their dominance in the automobile catalyst industry. Figure 1 illustrates the metal's industrial lifecycle, with the initial stage resulting in the discarding of tailings, with metal residuals varying based on ore and extraction technology. Historical mining tailings can become modern-day metal concentrations, while subsequent processing stages produce more complex and lower-volume slags. Product manufacturing losses are typically captured and recycled, with recovery opportunities in tailings and discarded metal-containing products. Tailings have low metal concentrations and are widely dispersed geographically.

Chemical methods such as excavation, precipitation, heat treatment, electroremediation, and chemical leaching for heavy metals (HMs) decontamination are costly and influenced by pollutant and soil characteristics (Nedjimi and Daoud, 2009a). These techniques have major disadvantages and drawbacks, including modification of soil properties, risk of soil fertility loss, small-scale application, and by-product generation (Nedjimi, 2021). In contrast, phytoremediation is a green approach that utilizes hyperaccumulator plants and their rhizospheric microorganisms to stabilize, transfer, or degrade pollutants in soil, water, and the environment. This technology is deemed efficient, cost-effective, and adaptable to the environment (Gupta and Chatterjee, 2014; Liu *et al.*, 2020a). Five types of phytoremediation have been applied based on soil conditions, pollutants, and plant species: phytodegradation, phytofiltration, phytoextraction, phytostabilization, and phytovolatilization (Ashraf, 2010; Nedjimi, 2020). Plants are classified as tolerant or hyperaccumulators of HMs if they show rapid growth, high biomass, and can extract and accumulate high amounts without toxicity (Gupta, 2002; Krämer, 2010). Thus, green technology has the potential to significantly aid in the remediation of HM-contaminated soils and agro-ecosystems (Bernard *et al.*, 2004). Plant hyperaccumulators have received greater attention in recent decades due to their potential for HM accumulation. However, there are some limitations for these plants to become efficient on a large scale. These limitations need to be overcome by transgenic approach applications to improve HM tolerance/accumulation of these plants (Rai *et al.*, 2020). The lifecycle of an element can lead to opportunities for green technologies for element recovery, promoting a circular economy, and reducing reliance on element extraction and purification. Phytoextraction is a global technology utilizing plants to extract metal from soil, aiming to mitigate hazards at industrial waste sites (Pilon-Smits, 2005). Recent years have seen extensive reviews of phytoextraction and its applications, including phytoremediation, phytodegredation, phytostabilization, phytovolatilization, and phytomining (Kukreja & Goutam, 2012). The 53 elements in the d-block are classified as "heavy metals" based on their density exceeding 5 g/cm^3 (Jarup, 2003) HMs, including cadmium (Cd), mercury (Hg), lead (Pb), arsenic (As), zinc (Zn), copper (Cu), nickel (Ni), and chromium (Cr), are metallic elements with high densities, atomic weights, and numbers (Gupta *et al.*, 2013). Angiosperms evolved with 19 essential elements for basic metabolism:

macronutrients (C, O, H, Mg, S, N, Ca, P, K) and micronutrients (Cu, Zn, Mn, Fe, Mo, B, Ni, Co, Cl, B). Besides, silicon (Si) is a beneficial element that plays a role in maintaining plant structures in some plants. Macro- and micronutrients are crucial in plant physiological and biochemical processes like chlorophyll biosynthesis, photosynthesis, DNA synthesis, protein modifications, redox reactions, sugar metabolism, and nitrogen fixation. In addition, zinc acts as a cofactor for over 300 enzymes and 200 transcription factors involved in membrane integrity maintenance, auxin metabolism, and reproduction (Ernst, 2006; Epstein, 1999; Prasad, 2012; Ricachenevsky *et al.*, 2013). Despite this, when they exceed their threshold concentrations, they are considered toxic to plant development, and their classification primarily depends on their density. Pb is a highly toxic HM with a soil retention time of 150–5,000 years and is reported to maintain high concentrations for up to 150 years (Yang *et al.*, 2005). Extensive agriculture, industrial, and mining activities are the primary sources of environmental contamination in Latin American countries, Europe, Africa, and China (Gong *et al.*, 2020; Chmielowska-Bąk & Deckert, 2021). Indeed, Mexico has been a profitable mining country since colonial times, producing silver, gold, copper, and lead, among other economically important metals (Mendoza-Hernández *et al.*, 2019). Recent research has examined endemic, native, and invasive plants that can adapt to mine tailings, such as copper concentration in ancient mining sites in Nacozari, Sonora, in Mexico (Santos-Francés *et al.*, 2017). The land use and cover area frame statistical topsoil survey (LUCAS) revealed high concentrations of Cd, Pb, Hg, and As in various European industrious regions, including Southern Saxony, Ruhr, Southern Poland, Central Rumania, Northern Spain, Lyon, and Nimes (Toth *et al.*, 2016). Contamination of water and soil, particularly in Africa, is causing widespread metal accumulation in crops, fish, cattle, and humans due to the presence of Cd and Pb contaminants (Yabe *et al.*, 2010). Metal levels in staple crops and vegetables (rice and maize), including As, Cd, Cr, and Pb, have been detected, sometimes exceeding permissible norms by 10–15 times (Sharma & Nagpal, 2020). Most articles focus on metal accumulation, toxicity, and plants' defense response, while there are limited studies on post-stress recovery (Afonne & Ifebida, 2020). Metal stress is less transient than other stresses, but soil metal concentrations may decrease over time due to uptake, erosion, or leaching, allowing plants to enter a post-metal-stress stage (Boyd, 2009). Stress tolerance evaluation requires examination of events following stress attenuation, as some plant species have adapted a quiescence strategy, halting growth during stress to conserve energy for restoration. In comparison, some species grow normally but may deplete energy and nutrient supplies, making tolerance estimation methods inexact. It's suggested to include recovery phases in tolerance tests (Striker *et al.*, 2011; Kumar *et al.*, 2019). Studies are currently focusing on "omic" tools, including ionomics (trace elements), metabolomics, transcriptomics, and proteomics, to address the problems. Further, it can enhance stress tolerance and aid in breeding and engineering programs for developing plants with desired agronomical traits (Lee *et al.*, 2007; Atkinson & Urwin, 2012). This review explores the potential of living plants to recover metals from mining industry waste, focusing on their potential for functional use in higher-value chemical applications. This also emphasizes the technology of phytoextraction, a green chemistry approach, which can be used to recover metals from plants, demonstrating its potential for creating novel, sustainable catalysts.

2. Metal Toxicity in Plants

Plants, being sessile organisms, are susceptible to environmental changes, including HM exposure, requiring them to develop strategies to cope with these negative consequences. The molecular and physiological mechanisms of metal toxicity and its impact on plants are well documented (Hossain *et al.*, 2012). Plants responses to external stimuli, such as HM toxicity, through various mechanisms, including stress signals, are detected and transmitted into cells, where they are then regulated to counteract their negative effects by affecting the cell's physiological, biochemical, and molecular status (Singh *et al.*, 2016). HM stress exposure to plants results in challenges in measuring sensing and signal transduction changes at the plant level. Monitoring early responses to stress, such as oxidative stress, transcriptomic and proteomic changes, and metabolite accumulation, can aid in studying sensing and signal transduction changes in plants. Tamás *et al.* (2010) found that early signs of metal toxicity in barley were similar to water deficiency signs, leading to overexpression of genes related to dehydration stress. Hernández *et al.* (2012) identified oxidative stress and glutathione depletion in alfalfa roots as early signs of sensing and signal transduction following HM exposure. Zhang *et al.* (2002) study revealed that high concentrations of As inhibited wheat seed germination and seedling growth. Imran *et al.* (2013)

found that exposure to As reduced the plumule and radicle length of *Helianthus annuus* L. seedlings. Moreover, As has been linked to decreased photosynthetic pigment, chloroplast membrane damage, and decreased enzyme activity due to protein sulfhydryl reactions, affecting nutrient balance and protein metabolism (Gupta and Chatterjee, 2017). HMs cause toxic effects in plants through four mechanisms: competition with nutrient cations for absorption at the root surface, similar to As and Cd. Metals/metalloids can bind with protein functional groups (sulfhydryl group (-SH), imidazole, and carboxyl groups), altering conformation and functioning and affecting protein folding and refolding processes, leading to misfolded or non-functional proteins (Tamás *et al.*, 2014). The displacement of essential cations from specific binding sites results in a function collapse. Metals can replace other elements in biomolecules, as seen in the case of replacing Ca in radish calmodulin with Cd, resulting in reduced activity. Several metals were found to substitute Mg in chlorophyll, causing a negative impact on the process of photosynthesis. Metal/metalloid toxicity indirectly involves reactive oxygen species overaccumulation, leading to oxidative stress, membrane damage, electrolyte leakage, an increased mutation rate, and reduced metabolic efficiency. The production of reactive oxygen species (ROS) can lead to the damage of macromolecules, as noted by Sharma & Dietz (2009) and DalCorso *et al.* (2013). The roots of sessile plants are the primary organs exposed to HMs, prompting extensive research to evaluate the stressor's impact (Keunen *et al.*, 2011). Eventually, HM exposure suppresses root growth in several plant species due to a decrease in mitotic activity, which is a crucial process for cell division and elongation. Liu *et al.* (1992) found Cr (VI) toxic to cell division, while Sundaramoorthy *et al.* (2010) observed that it causes cell cycle extension, inhibiting division and reducing root growth. Pena *et al.* (2012) found that Cd toxicity affects cell cycle progression through the S phase by decreasing cyclin-dependent kinase CDK expression, suggesting ROS may be involved. Yuan *et al.* (2013) found that excess Cu affects elongation and meristem zones by altering auxin distribution through (PIN1) PINFORMED1 protein, leading to Cu-mediated inhibition of primary root elongation. Peto *et al.* (2011) found that excess Cu can inhibit root length and morphology by altering auxin levels, which in turn counteracts nitric oxide function. Root growth inhibition leads to increased root diameter, suggesting that the plant cytoskeleton may be a potential target for HM toxicity (Zobel *et al.*, 2007). Therefore, according to studies, HMs may inhibit root growth, alter water balance and nutrient absorption, negatively impact shoot growth, and decrease biomass accumulation (Gupta 2013). Roots use mechanisms such as callose synthesis and deposition to prevent HM toxicity by creating barriers and enhancing root anatomy's plasticity. Roots not only prevent HM entry but also facilitate their transportation to aboveground plant parts, particularly in metallophytes or hyperaccumulator plants to sequester them in vacuoles, rendering them non-reactive. Plasma membranes (PMs) regulate cell entry and protection from stressors. *Arabidopsis halleri* and *arenosa* are more tolerant to HM stress due to low membrane depolarization, suggesting rapid membrane voltage changes could monitor metal toxicity (Kenderešová *et al.*, 2012). HMs alter cell metabolism, reducing growth and biomass accumulation, and may cause stunted stem and root length, stunted leaves, and chlorosis in younger and older leaves (Nagajyoti *et al.*, 2010; Gangwar *et al.*, 2011). Subsequently, metal toxicity impacts plants by altering key physiological and biochemical processes, blocking metabolic groups, and disrupting hormonal balance, nutrient assimilation, protein synthesis, and DNA replication (Hossain *et al.*, 2012; Hossain *et al.*, 2023; Helaoui *et al.*, 2023). Tomato seedlings have been found to experience severe negative impacts on photosynthetic indices such as photosynthetic rate and intracellular CO_2 concentration under Cd stress (Dong *et al.*, 2006). Maleva *et al.* (2012) found that Mn, Cu, Cd, Zn, and Ni significantly reduced chlorophyll content and photochemical efficiency in *Elodea densa*. Li *et al.* (2012) found that HMs such as Cu, Zn, Pb, and Cd negatively impact chlorophyll and carotenoids levels and PS II quantum yield in *Thalassia hemprichii*. Excessive Co significantly decreased water potential and transpiration rate in cauliflower leaves compared to excess Cu or Cr, while diffusive resistance and relative water content increased (Chatterjee and Chatterjee, 2000). HMs, such as Zn, can decrease CO_2 assimilation by diminishing ribulose bisphosphate (RuBP) carboxylase activity or reacting with the thiol group of RUBISCO (Ribulose-1,5-bisphosphate carboxylase/oxygenase), as reported in *Phaseolus vulgaris*. Muthuchelian *et al.* (2001) studied *Erythrina variegate* and found decreased RUBISCO activity under Cd stress, possibly due to the formation of mercaptide by Cd with a thiol group. Researchers found decreased CO_2 fixation due to decreased ATP and reductant pool, possibly due to Cd ions reducing the proton source for reduction reactions (Husaini and Rai, 1991; Ferretti *et al.*, 1995). Cu, a well-known inhibitor of carboxylase and oxygenase activities of RUBISCO, was found to decrease its activity in

Chenopodium rubrum by interacting with its essential cysteine residue (Schäfer *et al.*, 1992). Indeed, such stress induces ultrastructural changes in membranes, leading to reduced pigments, photosynthetic rate, PS II quantum yield, gas exchange, stomatal conductance, and CO_2 assimilation in various studies. Moreover, metals impact nitrogen metabolism, regulating plant growth and development, and enhance protease activity, affecting plant function and resource allocation (Chaffei *et al.*, 2004). In turn, the process reduces the activity of the enzymes nitrate reductase (NR), nitrite reductase (NiR), glutamine synthetase (GS), glutamine oxoglutarate aminotransferase (GOGAT), and glutamate dehydrogenase (GDH) involved in nitrate and ammonia assimilation, respectively. HM Cd impacts nitrogen metabolism by inhibiting nitrate uptake, transportation, nitrate reductase, and GS activity, thereby affecting primary N assimilation processes (Hernández *et al.*, 1997; Lea and Miflin, 2003). HMs, such as As, alter hormonal balance in plants, leading to their toxicities by altering auxin levels and around 69 microRNA expressions in *Brassica juncea* (Srivastava *et al.*, 2013). Hence, metallophytes, or hyperaccumulators, are unique in their ability to absorb large amounts of HMs from soil, making them useful in biogeochemical, biogeobotanical prospection, and phytoremediation technologies. Hyperaccumulators absorb HMs from soil, transferring them to shoots and accumulating them in aboveground organs at concentrations 100–1,000 times higher than non-hyperaccumulating species. However, hyperaccumulators, characterized by three hallmarks, do not cause toxic effects on plants due to their high concentration, according to recent advancements in understanding their mechanisms. HM tolerance strategies in *Thlaspi caerulescens* and *A. halleri* model plants have been studied for their ability to uptake, translocate, detoxify, and sequester HMs. Studies show hyperaccumulation is not caused by a novel gene but by differential expression of common genes between hyperaccumulators and non-hyperaccumulators (Verbruggen *et al.*, 2009).

3. Heavy Metal Uptake

The hyperaccumulation of HMs involves three intricate phenomena (Figure 2). They are remarkable at absorbing HMs from soil, but their uptake is influenced by factors such as pH, water content, and organic substances (Ma *et al.*, 2001; Xiao-e *et al.*, 2002). Recent data indicate that

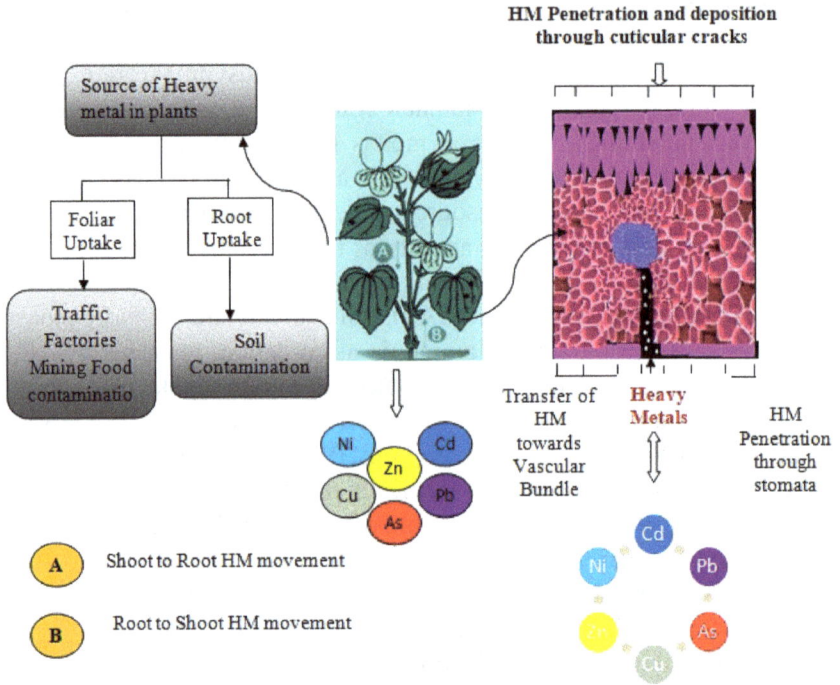

Figure 2.   Heavy metal uptake by foliar and roots in plants (adapted from Hasan *et al.*, 2019).

around 700 plant species out of the 300,000 vascular plants can undergo metal hyperaccumulation (Reeves, 2000). Further, HM uptake in plants requires a suitable transport system, with pH affecting proton secretion by roots, acidifying the rhizosphere, and enhancing metal dissolution and growth of metal-accumulating plant species (Bernal *et al.*, 1994; Peng *et al.*, 2006; Kuriakose & Prasad, 2008). Apart from pH, organic substances released from roots in hyperaccumulating plants affect growth, influencing HM solubility and absorption (Krishnamurti *et al.*, 1997; Peng *et al.*, 2006). The study by Whiting *et al.* (2000) suggests that increased root growth is also linked to certain plant species. Furthermore, comparative studies have shown that constitutive overexpression of genes in hyperaccumulating *Arabidopsis halleri* and *Thlapsi caerulescens* species can enhance HM uptake. Studies on *T. caerulescens* and *A. halleri* show increased Zn uptake due to overexpression of zinc-regulated transporter iron-regulated transporter protein (ZIP) genes, which encode plasma membrane-located transporters (Assunção *et al.*, 2001). The study found that ZTN1 and ZTN2 in *T. caerulescens* and ZIP6 and ZIP9 in *A. halleri* decreased Cd uptake with increasing Zn concentration, primarily due to the Zn transporter's preference (Weber *et al.*, 2006). Evidence suggests that P enters plant cells through phosphate transporters (Poirier & Bucher, 2002). A study on *Pteris vittata* and *Pteris tremula* revealed a higher density of phosphate/arsenate transporters in the root cell plasma membranes of *P. vittata* than *P. tremula*, likely due to overexpression of the constitutive gene. A study on Se hyperaccumulators *Astragalus bisulcatus* and *Stanleya pinnata* found higher Se/S ratios in shoots, indicating enhanced Se uptake through sulfate transporters (Galeas *et al.*, 2007). Hyperaccumulator plants, unlike non-hyperaccumulator plants, do not retain HM absorbed from roots but translocate it into shoots via xylem, involving various proteins. The proteins involved include HM-transporting ATPases, natural resistance-associated macrophage proteins (Nramp), Cation diffusion facilitator (CDF) family proteins, zinc–iron permease (ZIP) family proteins, and multidrug and toxin efflux (MATE) protein family. CPx-type and P1B-type ATPases transport toxic metals such as Cd, Pb, Cu, and Zn and regulate metal homeostasis and tolerance, respectively, using ATP across cell membranes (Axelsen & Palmgren, 1998; Williams *et al.*, 2000). HM ATPases (HMAs) overexpressed in the roots and shoots of hyperaccumulators indicate their upregulation compared to non-hyperaccumulators (Papoyan & Kochian, 2004). Nramp genes, involved in transporting HM ions, are found in the rice proteins OsNramp1, OsNramp2, and OsNramp3, which transport distinct but related ions in rice tissues (Belouchi *et al.*, 1997). CDF proteins transport Zn, Co, and Cd, regulating cation efflux out of the cytoplasmic compartment, and are also known as "cation efflux transporters" (Mäser *et al.*, 2001). ZNT1 and ZAT1 are Zn transporters from *T. caerulescens* and *Arabidopsis*, found in high levels in roots and shoots, respectively, belonging to the ZIP gene family (Van der Zaal *et al.*, 1999). The transporter protein MATE, specifically FDR3, is involved in HM translocation in *T. caerulescens* and *A. halleri* roots, with the gene encoding this protein playing a crucial role (Talke *et al.*, 2006; Van De Mortel *et al.*, 2006; Kraemer *et al.*, 2007). Therefore, all these studies provide strong evidence that multiple transporter proteins are involved in the translocation of HMs.

4. Detoxification/Sequestration of Heavy Metals

Hyperaccumulators are closely related plant species with unique abilities to bioaccumulate metals in their aerial organs more than other flora in the same location. Hyperaccumulators, after translocating, sequester and detoxify HMs, enabling them to survive in metal-contaminated areas without any toxic effects (Rascio & Navari-Izzo, 2011). Plants undergo detoxification/sequestration in their vacuoles (Figure 3), involving ATP-binding cassette (ABC), cation diffusion facilitators (CDF), HMA, and Nramp transporter families, as per various studies. ABC transporters, including HMT1, transport HMs into the vacuole, with MRP and PDR active subfamilies (Mendoza-Cózatl *et al.*, 2011). HMT1 in *Schizosaccharomyces pombe* aids in PC-Cd (phytochelatins–cadmium) complex transport (Ortiz *et al.*, 1992, 1995; Kuriakose and Prasad, 2008). A functional HMT1 homolog has been found in *Caenorhabditis elegans*

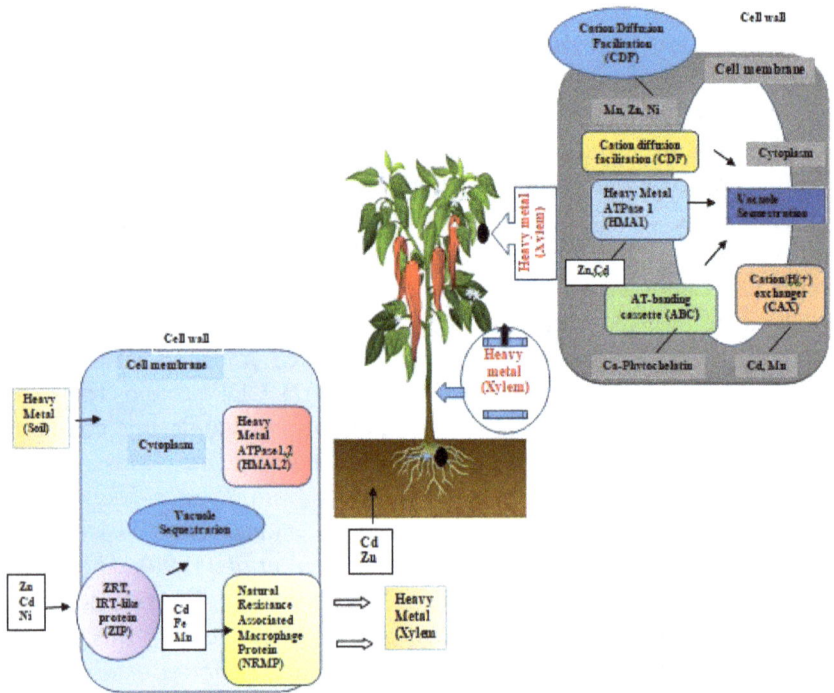

Figure 3. Detoxification/sequestration of heavy metals (adapted from Dubey *et al.*, 2018).

and *Drosophila*, but no such homolog has been studied in plants. *A. thaliana* transporters (AtMRP1 and AtMRP2) by transporting PC–Cd complexes into the vacuoles confer metal tolerance (Lu *et al.*, 1997, 1998). The CDFs transporter family, also known as metal tolerance protein (MTP), transports metal cations such as Zn^{2+}, Cd^{2+}, Co^{2+}, Ni^{2+}, or Mn^{2+} from the cytosol to the vacuoles (Kraemer *et al.*, 2007; Montanini *et al.*, 2007). Studies on *A. halleri* and *T. caerulescens* reveal higher expression of MTP1, MTP8, and MTP11 compared to non-hyperaccumulators (Becher *et al.*, 2004; Talke *et al.*, 2006; Vvan De Mortel *et al.*, 2006). Similarly, AhMTP1 protein, MTP11, and MTP8 are homologs of *Stylosanthes hamata* (ShMTP8), confirming Mn tolerance in *A. thaliana*, suggesting their role in metal tolerance (Delhaize *et al.*, 2003). In addition, HMA and Nramp transport metals from cytosol to the vacuoles, with HMAs potentially involved in detoxification mechanisms due to overexpression in *A. thaliana*. Aside from organic acids, in addition to transporter proteins, play a crucial role in detoxification processes by entrapping and chelating metal ions. Citrate binds with Ni in *Thlaspi goesingense* leaves, forming a metal – organic acid complex for chelation (Kramer *et al.*, 2000). Correspondingly, malate binds with Zn in *A. halleri* and Cd in *T. caerulescens*, as reported by Salt *et al.* (1999) and Sarret *et al.* (2002). Amino acids play a crucial role in hyperaccumulators by forming stable complexes with bivalent cations, thereby aiding in the sequestration of metal cations (Callahan *et al.*, 2006). For instance, histidine is involved in Ni hyperaccumulation, with a high concentration reported in the roots of Ni hyperaccumulators (Assuncao *et al.*, 2003). The overexpression of antioxidant-related genes, such as reduced glutathione (GSH), cysteine, and O-acetylserine, plays a crucial role in HM detoxification in hyperaccumulators (Anjum *et al.*, 2014). Studies show that upstream signaling of salicylic acid leads to increased serine acetyltransferase (SAT) activity and higher GSH levels (Freeman *et al.*, 2005). Overexpression of NgSAT in *Noccaea goesingense* led to increased GSH levels, resulting in increased Ni, Co, Zn, and Cd tolerance (Freeman *et al.*, 2004; Freeman and Salt, 2007).

Zeid *et al.* (2013) found that pretreatment of HMs solutions with precipitation and Ethylenediaminetetraacetic acid (EDTA) reduced cobalt toxicity in *Medicago sativa*, reducing growth and metabolic activities. Li *et al.* (2008) conducted an experiment on copper-stressed *Arabidopsis thaliana*, demonstrating that silicon can alleviate Cu toxicity. Si reduced leaf chlorosis, increased root-shoot biomass, and reduced stress-induced enzyme (phenylalanine ammonia-lyase) in plants. Si reduced the RNA levels of *Arabidopsis* copper transporter genes, specifically copper transporter 1 (COPT1) and HM ATPase subunit 5 (HMA5). Accordingly, Si enhances plant resistance to Cu toxicity, with limiting exposure to *Juglans regia, Robinia pseudoacacia, Eucalyptus* sp., and *Populus* sp. plantations alleviating Mn and Cu toxicity (Chatzistathis *et al.*, 2015). Hajiboland *et al.* (2013) found aluminum reduces Fe toxicity in tea plants, while Dufey *et al.* (2014) found Si reduces Fe-generated toxicities in rice plants. Rogalla & Romheld (2002) investigated the toxic effects of Mn in *Cucumis sativus*, revealing that decreasing Mn in intercellular washing fluid [barium chloride ($BaCl_2$) and diethylenetriaminepentaacetic acid (DTPA)] reduced stresses. Dragišić Maksimović *et al.* (2012) conducted a study on cucumber, revealing that Si can mitigate manganese-mediated toxicities in plants, as noted by Liang *et al.* (2007). Kumchai *et al.* (2013) investigated the potential of proline to partially mitigate molybdenum-induced stress in cabbage seedlings. Ali *et al.* (2015) found that gibberellic acid, a plant hormone, can alleviate Ni-induced stress in mungbeen plants, enhancing growth and yield. Likewise, jasmonic acid, a phytohormone, has been found to improve plant growth by reducing Ni-mediated toxicity in *Glycine max*, according to a study by Sarhindi *et al.* (2015). Siddiqui *et al.* (2013) highlighted the beneficial role of salicylic acid and nitric oxide in reducing Ni stress in wheat. Kaya *et al.* (2009) found that the application of Si (1.0 mM) in maize plants grown with high zinc concentrations improved growth, chlorophyll content, and relative water content. Likewise, researchers used 24-epibrassinolide, a plant stress hormone, to alleviate Zn-induced oxidative stress in radish seedlings, activating the antioxidative enzymatic system (Ramakrishna & Rao, 2012). *Berkheya coddii* is a drought-tolerant, fast-growing plant found in tropical Africa, particularly South Africa, known for its high biomass production and field-tolerant phytomining magnet. As one of over 532 globally known nickel hyperaccumulators, it has been studied for its efficacy in phytomining, with a pot trial showing a dry biomass of 22 t/ha (Robinson *et al.*, 1997; Mesjasz-Przybyłowicz *et al.*, 2004). Early phytomining trials focused on *Odontarrhena* plants, specifically muralis and corsica, which are annual and perennial flowering plants with high nickel hyperaccumulators. These hyperaccumulators are found in serpentine soils in southern Europe, northern Africa, and Asia Minor, with the highest species in the Mediterranean (Li *et al.*, 2003). *Hybanthus floribundus*, a woody shrub in southern and western Australia, has the potential to store nickel and cobalt up to 5,000 and 100 µg/g, making it a potential phytomining candidate due to its ability to thrive on mine tailings (Severne & Brooks, 1972). *Cannabis sativa*, an annual herbaceous plant in central and southern Asia, is a potential candidate for phytoextraction due to its high metal tolerance and biomass status. Its longer root allows for nickel extraction of 10 tons per hectare per year (Mura *et al.*, 2004). *Streptanthus polygaloides*, a mustard plant found in California's Sierra Nevada, has been discovered to hyperaccumulate nickel. Researchers conducted a phytoextraction study, revealing the plant can recover over 100 kg of nickel ha^{-1} from serpentine soils, potentially yielding $513 ha^{-1} (Boyd *et al.*, 2009). *Acacia*s, native to tropical and subtropical regions, are hyperaccumulators with high biomass and metal tolerance. Despite their longer growth time, they are economically viable due to their extensive root system, ability to grow in poor soils, and low input. *Acacia longifolia* and *Acacia sieberiana*, indigenous South African plants, are cyanogenic, producing organic thiocyanate in their tissues, aiding phytoextraction activities in soil (Abbas *et al.*, 2016; Khalid *et al.*, 2017).

5. Mechanisms of Phytoremediation

Berti & Cunningham (2000) describe a plant-based method that uses plants to extract and remove elemental pollutants or reduce their bioavailability in soil. Plants absorb ionic compounds in soil, establishing a rhizosphere ecosystem to accumulate HMs, reclaim polluted soil, and stabilize soil fertility (Ali *et al.*, 2013; Jacob *et al.*, 2018; DalCorso *et al.*, 2019). Phytoremediation, an autotrophic system powered by solar energy, offers numerous advantages, such as the following: (i) it is economically feasible, simple to manage, and cost-effective for installation and maintenance; (ii) eco-friendly practices can minimize the exposure of pollutants to the environment and ecosystem; (iii) applicability: it can be applied on a large scale and can be easily disposed of; (iv) it prevents erosion and metal leaching by stabilizing HMs, reducing the risk of contaminants spreading; (v) soil fertility can be enhanced by releasing organic matter into the soil, as suggested by various studies (Aken *et al.*, 2009; Wuana & Okieimen, 2011; Jacob *et al.*, 2018). Over the years, numerous studies have been conducted to understand the molecular mechanisms of HM tolerance and develop techniques to enhance phytoremediation efficiency. Some of the plants used in the phytoremediation process are listed in Table 1.

Table 1.   List of some plants used in phytoremediation.

S. No.	Plant	Metal	Phytoremediation	References
1	*Agrotis tenuis*	Lead	Phytostabilization	Anguilano et al. (2020)
2	*Alyssuim wulenianum*	Nickle	Phytoextraction	Reeves and Brooks (1983)
3	*Azolla pinnata, lemna minor*	Copper and Chromium	Phytoextraction	Jain et al. (1990)
4	*Alyssum lesbiacum*	Copper	Phytoextraction	Singer et al. (2007)
3	*Alyssum murale*	Lead	Phytoextraction	Bani et al. (2007)
4	*Arabidopsis thaliana*	Chromium Cadmium/Zinc	Phytoextraction	Saraswat and Rai (2009)
5	*Arabidopsis hallerii*	Cadmium	Phytoextraction	Bert et al. (2003)
6	*Amanita muscaria*	Mercury	Phytoextraction	Falandysz et al. (2003)
7	*Arobis gemmifera*	Cadmium and Zinc	Phytoextraction	Kubota and Takenaka (2003)
8	*Astragalus racemosus*	Selenium	Phytoextraction	DeTar et al. (2015)
9	*Brassica juncea*	Lead, Copper, Nickel	Phytoextraction	Ebba and Kochian (1997)
10	*Astragulus bisukatus, Brassica juncea*	Selenium	Phytoextraction	Ellis et al. (2004)
11	*Brassica oleracea*	Zinc	Phytoextraction	Ebbs, SD; Kochian, L. V. (1997)
12	*Brassica napus*	Cadmium	Phytoextraction	Selvam and Wong (2008)
13	*Crotalaria juncea*	Nickle and chromium	Phytoextraction	Saraswat and Rai (2009)
14	*Chengiopanax sciadopkvltoides*	Manganese	Phytomining	Mizuno et al. (2008)
15	*Festuca arundinacea*	Boron	Phytovolatiization	Steliga and Kluk (2020)
16	*Haumaniastrum robertii*	Cobolt	Phytoextraction	Van Der Ent et al. (2019)
17	*Helianthus annus*	Uranium	Rhizofilteration	Lee et al. (2010)

18	*Hibiscus cannibus*	Boron	Phytovolatilization	Vishnoi et al. (2017)
19	*Ipomea alpine*	Copper	Phytoextraction	Napoli et al. (2019)
20	*Lemn agibba*	Arsenic	Phytoextraction	Mkandavire and Dude (2005)
21	*Lotus corniculatus*	Boron	Phytovolatilization	Garbisu et al. (2002)
22	*Pteris vittata*	Copper, Nickel, Zinc	Phytoextraction	Ma et al. (2001)
23	*Psychotria douarrei*	Nickel	Phytoextraction	Davis et al. (2001)
24	*Pelargonium sp.*	Cadmium	Phytoextraction	Dan et al. (2002)
25	*Pistca stratiotes*	Silver, Cadmium, Chromium, Copper, Nickel and Lead	Phytoextraction	Odjegba and Fasidi (2004)
26	*Piptathertan miliacetall*	Lead	Phytoextraction	Garcia et al. (2004)
27	*Raphanus sativus*	Cadmium	Phytoextraction	
28	*Rorippaglobosa*	Cadmium	Phytoextraction	Sun et al. (2010)
29	*Sebertia acuminate*	Nickel	Phytoextraction	Jaffré et al. (1976)
30	*Sedum alfredii*	Lead	Phytoextraction	Xiong et al. (2004); Banasova and Horak (2008)
31	*Sebania drummondi*	Lead	Phytoextraction	Sharma et al. (2004)
32	*Silense vulgais*	Zinc	Phytoextraction	Ramteke et al. (2008)
33	*Tamarix smyrnensis*	Cadmium	Phytoextraction	Manousaki et al. (2008)
34	*Thalaspi caerulescens*	Cadmium	Phytoextraction	Zhao et al. (2003)
35	*Thlaspi caerulescens*	Cadmium Nickel	Phytoextraction	Assunção et al. (2003)

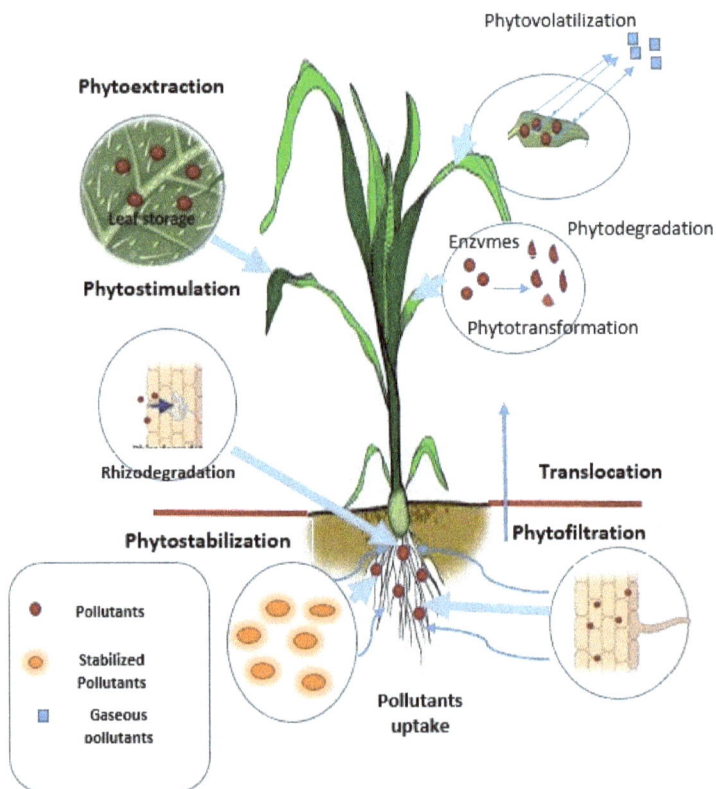

Figure 4.   Mechanism of phytoremediation (adapted from Krishnasamy *et al.*, 2021).

There are five categories of phytoremediation technologies (Figure 4) used in soil cleanup, as described in the following.

5.1 *Phytoextraction*

Phytoextraction is a method used to plant species that accumulate the highest amount of pollutants in their shoots, exceeding 0.1% of DW. The selection criteria for these plants can be determined by the degree of contaminant translocation from roots to shoot (Susarla *et al.*, 2002; Rehman *et al.*, 2019). The strategy employs two methods: continuous phytoextraction and induced phytoextraction, with continuous phytoextraction utilizing endemic plants with natural high HMs accumulation capabilities. The addition of chemical substances like chelates has been shown to enhance the accumulation of plant metals to induce phytoextraction (McGrath *et al.*, 2002). Phytoextraction is enhanced by plants with a high growth rate and deeper root systems. Begonia *et al.*'s (2002) study demonstrated that coffee weed (*Sesbania exaltata*) effectively removes Pb from contaminated soils. Fourati *et al.* (2016) found that Ni was accumulated at a higher level (1050 μg g^{-1} DW) in the aboveground part of *Sesuvium portulacastrum* (Fourati *et al.*, 2016). Jacobs *et al.* (2018) found that *Noccaea caerulescens* leaves had a Zn concentration exceeding 300 g Cd ha^{-1} two months before transplantation under field conditions. Ghazaryan *et al.* (2019) compared the effectiveness of *Melilotus ofcinalis* and *Amaranthus retrofexus* in treating contaminated soils with Cu and Mo. This study revealed that *A. retrofexus* prefers Cu and Mo accumulation in shoots, while *M. ofcinalis* prefers Zn storage in roots. Yang *et al.* (2020) study on Napier grass varieties revealed *that P. purpureum* cv. *Guiminyin* accumulates the highest amounts of Cd and Zn in their shoots. Khalid *et al.*'s (2020) study found that *Alternanthera bettzickiana*, after eight weeks of treatments, accumulates twice as much Cu in its shoots compared to the control.

5.2 *Phytotransformation/Phytodegradation*

Phytodegradation, or phytotransformation, is the process of removing pollutants from plants through metabolic processes or externally through enzymes (dehalogenases, nitroreductases, and peroxidases) produced by roots (Schnoor *et al.*, 1995). Yellow poplar, genetically modified, can undergo a transformation from highly toxic mercury Hg^{2+} to a less toxic Hg^0 form in tissue culture with higher mercury concentrations (Rugh *et al.*, 1998). This method enables plants to convert pollutants into nonhazardous components. Study of Das *et al.* (2010) found that *Vetiveria zizanioides* plants can effectively remove 97% of trinitrotoluene (TNT) from the soil. Hannink *et al.* (2007) found that *Nicotiana tabacum* roots produce the NfsI nitroreductase enzyme, which contributes to the degradation of TNT. Just & Schnoor (2004) reaffirmed that *Populus deltoids* plants can convert RDX (Hexahydro-1,3,5-trinitro-1,3,5-triazine) into metabolic components. Certain plants offer optimal conditions for the association of bacteria and mycorrhizae to develop and degrade toxins. The degradation components are either volatilized or incorporated into the soil matrix (de Farias *et al.*, 2009). The plant's production of sugars and organic acids, which are released, has been found to increase the populations of fungi and bacteria (Shahzad *et al.*, 2015). Rhizodegradation can be enhanced by enhancing soil characteristics including aeration and moisture content (Kirk *et al.*, 2005). Papadopoulos & Zalidis conducted an experiment to treat contaminated wetland with terbuthylazine (TER) and discovered *Typha latifolia* rhizomes as a promising species for TER phytodegradation. The study showed *Rhizophora mangle* mangrove, a plant growth-promoting rhizobacterium, can degrade polycyclic aromatic hydrocarbons (PAHs) in contaminated sediment (Sampaio *et al.*, 2019).

5.3 Phytovolatilization

Phytovolatilization is the process of assimilating pollutants, transferring them to the shoot, and then volatilizing them in the atmosphere through stomatal leaves (Tollsten & Muller, 1996). Bañuelos *et al.* (1997) suggested that Indian mustard (*Brassica napus*) can effectively perform the phytovolatilization of Se from soil. Bizily *et al.* (1999) found that *Arabidopsis thaliana* converts organic Hg salts to volatile elemental form, while *Brassica juncea* hydroponically removes up to 95% of Hg through phytovolatilization and plant accumulation (Moreno *et al.*, 2008). Ashraf *et al.* (2010) demonstrated that transgenic *Nicotiana tabacum*, carrying the merA gene, effectively removed Hg from soils contaminated with the substance.

5.4 Rhizofiltration

Rhizofiltration is a method used to remove pollutants from water and liquid waste by precipitating toxins onto surface roots or absorption of soluble pollutants into roots (Dushenkov *et al.*, 1995). This technique is significantly influenced by its fibrous root system and large surface areas of roots (Sampaio *et al.*, 2019). Plants that can efficiently absorb large amounts of soil water are suitable for this purpose (Susarla *et al.*, 2002). Amaya-Chavez *et al.*'s (2006) study demonstrates that *Typha latifolia* can effectively remove methyl parathion (MeP) from hydromorphic soils. Yang *et al.* (2015) discovered that the bean species, *Phaseolus vulgaris*, effectively removes uranium and cesium from groundwater. Oustriere *et al.* (2017) found that *Arundo donax* is an effective Poaceae for rhizofertilizing copper from constructed wetlands. Kodituwakku & Yatawara's

(2020) recent study found that *Eichhornia crassipes*, *Salvinia molesta*, and *Pistia stratiotes* are promising terrestrial plants for removing HMs from industrial sewage sludge.

5.5 *Phytostabilization*

This strategy involves the storage of pollutants by plant roots or the precipitation of pollutants from root exudates (Peer *et al.*, 2005). This method minimizes the movement of pollutants and prevents their migration to groundwater. Nedjimi & Daoud (2009a) found that *Atriplex halimus*, a local North African halophyte, is a promising candidate for phytostabilizing Cadmium, with high concentrations in roots. Plants suitable for phytostabilization have morphology and root depth, with numerous fine root system extents providing maximum soil matrix contact. Root depth significantly differs based on plant species, moisture, soil texture, calcareous encrust, dry conditions, and soil amendments. Some plants can alter metal solubility and mobility through root exudate excretion, while others sequester significant amounts of HMs in their roots. Al Chami *et al.* (2015) conducted pot experiments comparing the remedy effects of *Sorghum bicolor* and *Carthamus tinctorius* on Ni, Pb, and Zn. Furthermore, *Erica australis*, a phytostabilizer species, can uptake metals like Cu, Cd, and Pb without damage, indicating its potential for accumulating metals through its roots (Monaci *et al.*, 2019). Bacchetta *et al.* (2018) found that Zn, Cd, and Pb uptake is primarily limited in the root tissue of *Helichrysum microphyllum*, making it suitable for phytostabilization. Manzoor *et al.* (2020) found that *Stigmatocarpum criniforum* and *Pelargonium hortorum*, under Pb stress, *P. hortorum*'s aboveground part accumulates more Pb than its roots. *Quercus robur* and *Salix alba* are potential candidates for phytostabilizing Cd and Cu in polluted soils, with *Quercus robur* being the best accumulator in roots (Manzoor *et al.*, 2020; Mataruga *et al.*, 2020).

6. Limitations of Metal Recovery by Green Plants

Phytoextraction is a promising method for metal recovery, but it's crucial to consider its limitations and risks during operation design. The main concern is the slow growth rates of plants, leading to a lower production rate compared to mechanical methods (Singh *et al.*, 2003; Shah & Nongkynrih,

2007). Metal-hyperaccumulating plants' low biomass production can hinder large-scale site decontamination, relying on harvested biomass and metal accumulation by dry matter, except for Ni hyperaccumulator species (Cunningham & Ow, 1996). Limited biomass raises economic concerns as it may take time for crops to accumulate enough metal to recover initial planting and harvesting costs (Bradshaw, 1997). Soil environments with physical or chemical properties can limit the growth of all plant species, whether they are hyperaccumulators or non-hyperaccumulators. Tailings of fine particle size create a growth medium with limited porosity, oxygen content, and drainage potential, resulting in anoxic conditions that hinder plant growth (Schnoor *et al.*, 1995). Chemical impediments can be triggered by the presence of potentially phytotoxic elements like Cu, Cd, and As in mineralized ore. The high concentration of essential nutrient parameters in waste rock or tailings can negatively impact plant growth if not recovered during processing (Hunt *et al.*, 2013). However, mineralized rock's low organic C and N status is crucial for microbiological soil health and nutrient cycling, which are vital for plant growth (Macek *et al.*, 2000). Phytoremediation is a method that involves the use of plant roots to clean and restore soil, but its effectiveness is limited by the depth of the roots. Surface contaminants can only be removed through sequential stripping, allowing plants to access subsurface metals, but this process is likely to increase costs. Therefore, the study highlights the limitations, including the climatic and geological conditions of the polluted target area as well as the accessibility of the site. Van der Ent *et al.* (2013) reported the failure of large-scale phytomining in Sulawesi, Indonesia, due to the inability of the temperate species to thrive in tropical environments. Environmental concerns involve the possibility of metals entering trophic chains through herbivory of metal-containing biomass (Singh *et al.*, 2003). The treatment and disposal of harvested metal-containing biomass pose challenges as it must be handled as hazardous waste. Further, the introduction of non-native plant species to recovery sites may pose a threat to local biodiversity. Phytoextraction constraints extend beyond plants and the environment; resource issues must also be considered (Ghosh & Singh, 2005). Phytoextraction for industrial metal recovery requires resource security, and fluctuations in metal prices shift mine waste from waste to reserve (Anderson *et al.*, 1999). The gold price has increased since the late 1990s, indicating that tailings from a 1998 gold mine may be considered a resource in 2013 (*USGS Minerals Yearbook*, 2013; Anderson *et al.*, 1998). Improved process technology can potentially reclassify mine waste over time, impacting phytomining project development by securing resource access.

7. Utilization and Applications of Recovered Metals

Phytoextraction of metals offers three benefits: land decontamination, surface metal resource exploitation, and metal recovery from low metal concentration ore bodies. The concentration of metals in biomass, from below ground to above ground, presents potential for recovery and recycling for various applications (Corzo Remigio *et al.*, 2020). However, phytoextraction technologies can aid in recovering elements from marginal mining sites or tailing, promoting a circular economy and elemental sustainability, and often treating biomass as bio-ore (Nkrumah *et al.*, 2018). Research shows that certain plant-sequenced metals can be stored as pure metallic nanoparticles within biomass, providing a third key benefit to phytoextraction for metal uptake. Nanoparticles, ranging from 1 to 100 nm in size, possess unique properties such as large surface area to volume ratio, high catalytic efficiency, and strong adsorption ability (Hunt *et al.*, 2014). The extraction of plant-synthesized nanoparticles using methods such as freeze-thawing, biomass incineration, and chemical leaching is laborious, costly, and energy-intensive. In situ use of nanoparticles within biomass is gaining interest. Studies show plants synthesized Zn and Ni nanoparticles as catalysts for alcohol chlorination and Friedel-Crafts chemistry, but ashing may cause nanoparticle agglomeration and reduced activity. Losfeld *et al.* (2012) found *Thlaspi caerulescens* can hyperaccumulate zinc, producing a Lewis acid catalyst for organic synthesis which was efficient for the chlorination of alcohols and promising for further reuse. Parker *et al.* (2014) demonstrated that wild-type *Arabidopsis* plants can salvage palladium and produce plant-based palladium catalysts, which exhibit excellent catalytic activity compared to other non-conventional palladium catalysts. Harumain *et al.* (2017) studied the ability of liquid-culture-grown *Arabidopsis, mustard, miscanthus*, and 16 willow species to uptake palladium nanoparticles. They found a threshold concentration of palladium in dried tissues with catalytic activity comparable to commercially available 3% palladium on carbon-supported catalysts. Plant-synthesized nanoparticles offer a promising value-added application for phytomined metals in environmentally relevant catalytic reactions, utilizing their unique properties for reactions not possible or low-yielding using conventional catalysts. Sharma *et al.*'s (2007) study on *Sesbania drummondii* seedlings revealed Au nanoparticles as effective catalysts for pollution reduction, without extracting the nanoparticles from the plant. TiO_2 nanoparticles are effective catalysts for photocatalysis, dye-sensitized solar cells, and photovoltaic cells due to their large surface area. Plant-synthesized TiO_2

nanoparticles, formed through phytoextraction using water hyacinth, have shown potential for photocatalysis in syngas production and hydrogen production through photolytic dissociation of water (Mahmoud *et al.*, 2021). Nanoparticles like Pt, Pd, or Rh are commonly used in automobile catalytic converters for environmental catalysis, which have been used for nearly 40 years for exhaust gas scrubbers and emission control (Deka *et al.*, 2022). PGMs, a finite resource, can be exploited through phytoextraction by converting plants containing metals into heterogeneous catalysts. This method retains the activity of nanoparticles, providing a green method for the green utilization of high-value materials (Patil & Patil, 2021). Au nanoparticles are extensively researched for biomedical applications, including biosensors, geology, clinical chemistry, cancer cell photothermolysis, targeted drug delivery, and optical bioimaging (Huang *et al.*, 2007). Plant-synthesized nanoparticles could enhance safety in biomedical applications by eliminating toxic chemicals commonly used in conventional synthesis techniques, despite no reported examples of their use in biomedical applications (Simon *et al.*, 2022). Biomass containing phytoextracted metals, like Cu, can improve bio-oil quality through fast pyrolysis. Cu catalyzes thermo-decomposition, improving yield and heating value and preventing metal contamination due to its non-volatilizing nature (Liu *et al.*, 2012). Researchers have explored Se-containing plants for fortified foods, biofuels, bioherbicides, and green fertilizers. Bañuelos & Hanson (2010) showed increased strawberry yields when Se-enriched seed meals from canola and mustard plants were used. Soil treatment increased fruit nutrient content, and Se-enriched amendment reduced summer-germinating and winter annual weed emergence, suggesting potential applications in organic agriculture despite large-scale trials. The potential of nanoparticles from phyto-extracted biomass in higher-value applications is significant, but further green chemistry research is needed to fully realize this potential (Johar *et al.*, 2022).

8. Improving Phytoremediation through Biotechnological Approaches

8.1 *Genetic engineering*

Genetic engineering is a promising technique for improving plant phytoremediation against HM pollution by inserting a foreign gene from an organism into the target plant's genome (Yan *et al.*, 2020). Genetic engineering involves DNA recombination, transferring foreign genes to plants, allowing for desirable traits for phytoremediation, and the transfer of desirable genes to sexually incompatible species, a significant improvement over traditional breeding methods (Berken *et al.*, 2002). Therefore, genetic engineering for developing transgenic plants with desired traits offers promising phytoremediation prospects, particularly in fast-growing, high-biomass species with high tolerance and HM accumulation ability. Hence, fast-growing, high-biomass plants are engineered to enhance HM tolerance or increase accumulation ability, requiring genetic engineering genes based on plant knowledge of these mechanisms (Yadav *et al.*, 2010). HMs increase oxidative stress defense, leading to HM tolerance. Enhancing antioxidant activity, achieved by overexpressing antioxidant machinery genes, is a common strategy to increase tolerance. It often involves introducing and overexpressing genes involved in HM uptake, translocation, and sequestration to increase metal accumulation (Mani and Kumar, 2014; Das *et al.*, 2016). Genes encoding HM/metalloid transporters, such as ZIP, MTP, MATE, and HMA family members, can be transferred and overexpressed in target plants to enhance HM accumulation. Metal chelators, acting as metal-binding ligands, enhance HM bioavailability, promote uptake, and facilitate intracellular sequestration, making genetic engineering a promising strategy for increasing HM accumulation (Wu *et al.*, 2010). Morphology and root depth are crucial in selecting plant species for soil remediation. Shallow roots are ideal for surface-contaminated soils, while deep roots are suitable for deeper soils. Plants with large root volumes, high foliage biomass, high metal assimilation, and high exudate production are useful for phytoremediation, reducing soil remediation time (Lasat, 2002). Recent molecular tools improve our understanding of metal hyperaccumulation eco-physiology. Gene overexpression can reduce stress from HMs and enhance plant phytoremediation capacity (Yang *et al.*, 2005). Recent biotechnological advancements focus on gene expression to overcome limitations in phytoremediation processes. Mercuric ion reductase (merA) and organomercurial lyase (merB) are bacterial enzymes used to enhance Hg detoxification, converting organic Hg into the less toxic ionic form Hg^{2+} (Hui *et al.*, 2023). The SbMT-2 gene from *Salicornia brachiata* has been found to regulate ROS scavenging in transgenic *Nicotiana tabacum*, thereby reducing the tolerance to HMs (Cu, Zn, and Cd) (Chaturvedi *et al.*, 2023). Macrophage protein (Nramp) significantly contributes to HMs decontamination, as demonstrated by the overexpression of SaNramp6 in *Sedum alfredii*, enhancing Cd accumulation in transgenic *Arabidopsis thali* (Chen *et al.*, 2017). Cai *et al.* (2019) suggest rice OsHMA3 expression decreases shoot Cd accumulation in transgenic tobacco, while co-expression of OsLCT1, OsHMA2, and OsZIP3 transporters increases Cd and Zn uptake and translocation in *Oryza sativa* (Tian *et al.*, 2019). Khan *et al.* (2016) identified novel rice genes HPP (HM-associated plant protein) and HIPP (HM-associated isoprenylated plant protein) tolerant to Cu, Zn, Cd, and Mn, enhancing Cu tolerance and accumulation in transgenic tobacco. Liu *et al.* (2020a) discovered that the metallothionein PpMT2 gene, involved in HMs tolerance in *Physcomitrella patens*, could be utilized in transgenic *Arabidopsis* plants.

8.2. *Plant-associated microorganisms (rhizosperic microorganisms)*

Bioremediation involves using plant growth-promoting bacteria (PGPB) to colonize the rhizospheric system, stimulate plant growth and mineral nutrition, and potentially degrade toxic contaminants. PGPB enhances plant phytoremediation by allowing roots to uptake HMs, decontaminating soil with siderophores and organic acids, thereby enhancing the bioavailability of HMs by decreasing soil pH (Sirhindi *et al.*, 2015; Lopez *et al.*, 2022). Bacteria like polysaccharides and glomalin secreted by other bacteria contribute to the phytostabilization of HMs by reducing their mobility. PGPR significantly enhances phytoremediation processes by improving plant detoxification rates, enhancing enzymes, promoting pollutant degradation, and modifying soil pH, with many bacteria strains increasing plant HM tolerance (Wang *et al.*, 2022). *Arthrobacter* inoculated to *Ocimum gratis-simum* induces the phytoextraction of Cd by roots (Prapagdee & Khonsue, 2015). Guo and Chi (2014) demonstrated that PGRB *Bradyrhizobium* sp. can inhibit the growth and promote the uptake of Cd in *Lolium multiforum* and *Glycine max* seedlings in Cd-contaminated soil. Szuba *et al.* (2020) found that the Pb-tolerant *Paxillus involutus* strain can promote Pb tolerance in *Populus canescens* seedlings, while bacterial endophytes such as *Pantoea stewartii* ASI11 and Enterobacter sp. HU38 increase *Leptochloa fusca* plant phytostabilization in U and Pb-contaminated soils. *Mesorhizobium loti* HZ76 and the *Bacillus* XZM strain improve *Robinia pseudoacacia* growth and phytoremediation capacity, while *Vallisneria denseserrulata* plant's detoxification efficiency is significantly enhanced (Fan *et al.*, 2018). Arbuscular mycorrhizal fungi (AMF) decontaminate phosphorus in roots through two strategies: immobilization and phytoextraction. They produce chelating agents and adsorb HMs to fungal cell walls, improving plant growth and rhizosphere uptake (Cabral *et al.*, 2015; Shabani *et al.*, 2016). The inoculation of *Cassia italica* by AMF significantly improved the Cd tolerance by preventing its translocation to aerial parts (Raza *et al.*, 2020). *Festuca arundinacea* plants and *Glomus mosseae* fungi exhibit enhanced Ni translocation and expression of ABC transporter and metallothionein genes. AMF infection significantly improves maize growth, phosphorus pool, and HMs uptake in soil contaminated with Sr and Cd (Chang *et al.*, 2018). Armendariz *et al.* (2019) found soybean plants with *Bradyrhizobium japonicum* E109 and *Azospirillum brasilense* Az39 show better As stress tolerance, while the endophytic fungus *Piriformospora indica* improves sunflower seedlings' physiological status. Similarly, a recent study by Rahman *et al.* (2019) found that exogenous infection of *Artemisia annua* with *Piriformospora indica* can significantly increase its tolerance to As stress. Earthworms, also known as "ecosystem engineers," are crucial soil macro-invertebrates involved in organic matter decomposition, nutrient cycling, and soil condition improvement through their gut microfora, they secrete organic acids such as fulvic and humic acids, which decrease soil pH and improve nutrient and HM bioavailability in the rhizosphere (Sharma *et al.*, 2020). Wang *et al.* (2020) found that incorporating earthworms into the culture medium improves the phytoremediation capacity of Cadmium in *Solanum nigrum*. Bongoua-Devisme *et al.* (2019) found that *Pontoscolex corethrurus* can improve the Cr and Ni tolerance of *Acacia mangium*. The integration of *Rhizoglomus clarum* enhances the phytoextraction capacity of *Canavalia ensiformis* plants in sandy soil contaminated with Cu (Santana *et al.*, 2019). *Brassica juncea* plants with the *Esenia fetida* earthworm significantly improve Cd detoxification efficiency, while black oat plants' ability to remove Cd, Cr, and Pb is enhanced by vermicompost using *Ensenia Andrei* (Hoehne *et al.*, 2016; Kaur *et al.*, 2018). PGR-assisted phytoremediation improves HMs accumulation in plant tissues, identifying four beneficial plant hormones: auxins (IAA), cytokinins, gibberellins, and abscisic acid (ABA). Phytohormones enhance plant growth and tolerance to HMs, as they improve accumulation and HMs tolerance during early growth stages, helping plants escape toxicity (Zhu *et al.*, 2013). The addition of 0.05 M auxin has been found to significantly enhance the tolerance of Arabidopsis thaliana against Cd with minor damage. The application of 10 and 100 mM IAA in a nutrient solution can mitigate the harmful effects of Cd-stressed *Trigonella foenum-graecum* by inhibiting its uptake and regulating the ascorbate-glutathione cycle (Begonia *et al.*, 2002). Ji *et al.* (2015) found that the application of gibberellic acid 3, at concentrations of 10, 100, and 1000 mg/L, significantly enhances the biomass and phytoremediation efficiency of *Solanum nigrum*. Supplemental ABA enhanced growth in Zn-stressed *Vitis vinifera* by promoting the expression of ZIP and detoxification-related genes (Song *et al.*, 2019). Leng *et al.* (2020) found that ABA supplementation mitigates Cd adverse effects on mung bean growth by protecting membrane lipid peroxidation and modulating antioxidative defense systems (Ali *et al.*, 2015).

9. Conclusion and Future Outlook

Efficient metal recovery and recycling from contaminated biomass are crucial for green chemistry's economic potential, necessitating the development of innovative technologies for process valorization. Currently, only plants with Au, Cu, Se, Zn, and Ti are used in green chemistry. Expanding this range and applications is crucial, but the synthesis of active nanoparticles is feasible. Metal depletion and environmental contamination are growing issues. Life cycle assessments help understand metal flow and highlight activities causing significant metal losses. Lifecycle assessments can identify areas of concern and exploitation opportunities, focusing on phytoextraction's potential for metal recovery from newly defined resources. Phytoextraction is a crucial green remediation and metal acquisition method, with future research focusing on creating new hyperaccumulators and developing novel routes for metal uptake. Various bioremediation approaches, including genetic engineering, transgenic transformation, phytoremediation with phytohormones, microbes, AMF inoculation, and nanoparticle addition, are widely used to improve plant phytoremediation potential. Plants natural ability to form metal nanoparticles has not been fully utilized, offering potential for green chemistry to produce novel materials, catalysts, and chemicals. Research shows phytoextracted metals have limited applications, requiring further innovative research for the full potential of this green technology.

Acknowledgments

The authors sincerely acknowledge the respective institutional heads for their kind support in writing this review.

References

Abbas, A., Al-Amer, A. M., Laoui, T., Al-Marri, M. J., Nasser, M. S., Khraisheh, M., & Atieh, M. A. (2016). Heavy metal removal from aqueous solution by advanced carbon nanotubes: Critical review of adsorption applications. *Separation and Purification Technology, 157*, 141–161.

Afonne, O. J. & Ifebida, E. C. (2020). Heavy metals risks in plant foods — Need to step up precautionary measures. *Current Opinion in Toxicology, 22*, 1–6.

Al Chami, Z., Amer, N., Al Bitar, L., & Cavoski, I. (2015). Potential use of *Sorghum bicolor* and *Carthamus tinctorius* in phytoremediation of nickel, lead and zinc. *International Journal of Environmental Science and Technology, 12*, 3957–3970.

Ali, H., Khan, E., & Sajad, M. A. (2013). Phytoremediation of heavy metals— concepts and applications. *Chemosphere, 91*(7), 869–881.

Ali, M. A., Asghar, H. N., Khan, M. Y., Saleem, M., Naveed, M., & Niazi, N. K. (2015). Alleviation of nickel-induced stress in mungbean through application of gibberellic acid. *International Journal of Agriculture and Biology, 17*(5), 990–994.

Amaya-Chavez, A., Martinez-Tabche, L., Lopez-Lopez, E., & GalarMartinez, M. (2006). Methyl parathion toxicity to and removal efficiency by Typha latifolia in water and artifcial sediments. *Chemosphere, 63*, 1124–1129.

Anderson, C. W. N., Brooks, R. R., Chiarucci, A., LaCoste, C. J., Leblanc, M., Robinson, B. H., Simcock, R., & Stewart, R. B. (1999). Phytomining for nickel, thallium and gold. *Journal of Geochemical Exploration, 67*(1–3), 407–415.

Anderson, C. W., Brooks, R. R., Stewart, R. B., & Simcock, R. (1998). Harvesting a crop of gold in plants. *Nature, 395*(6702), 553–554.

Anjum, N. A., Umar, S., & Iqbal, M. (2014). Assessment of cadmium accumulation, toxicity, and tolerance in Brassicaceae and Fabaceae plants — Implications for phytoremediation. *Environmental Science and Pollution Research, 21*, 10286–10293.

Armendariz, A. L., Talano, M. A., Olmos Nicotra, M. F., Escudero, L., Breser, M. L., Porporatto C., & Agostini, E. (2019). Impact of double inoculation with *Bradyrhizobium japonicum* E109 and *Azospirillum brasilense* Az39 on soybean plants grown under arsenic stress. *Plant Physiology & Biochemistry, 138*, 26–35.

Ashraf, M., Ozturk, M., & Ahmad, M. S. A. (eds.) (2010) Toxins and their phytoremediation. In *Plant Adaptation and Phytoremediation*, pp. 1–32. Berlin: Springer.

Ashraf, S., Ali, Q., Zahir, Z. A., Ashraf, S., & Asghar, H. N. (2019). Phytoremediation: Environmentally sustainable way for reclamation of heavy metal polluted soils. *Ecotoxicology and Environmental Safety, 174,* 714–727.

Assunção, A. G., Schat, H., & Aarts, M. G. (2003). Thlaspi caerulescens, an attractive model species to study heavy metal hyperaccumulation in plants. *New Phytologist, 159*(2), 351–360.

Atkinson, N. J. & Urwin, P. E. (2012). The interaction of plant biotic and abiotic stresses: From genes to the field. *Journal of Experimental Botany, 63*(10), 3523–3543.

Axelsen, K. B. & Palmgren, M. G. (1998). Evolution of substrate specificities in the P-type ATPase superfamily. *Journal of Molecular Evolution, 46,* 84–101.

Bacchetta, G., Boi, M. E., Cappai, G., De Giudici, G., Piredda, M., & Porceddu, M. (2018). Metal tolerance capability of *Helichrysum microphyllum* Cambess. subsp. *tyrrhenicum* Bacch., Brullo & Giusso: A candidate for phytostabilization in abandoned mine sites. *Bulletin of Environmental Contamination and Toxicology, 101,* 758–765.

Bañuelos, G. S., Ajwa, H. A., Mackey, B., Wu, L. L., Cook, C., Akohoue, S., & Zambrzuski, S. (1997). Evaluation of different plant species used for phytoremediation of high soil selenium. *Journal of Environmental Quality, 26*(3), 639–646.

Bañuelos, G. S. & Hanson, B. D. (2010). Use of selenium-enriched mustard and canola seed meals as potential bioherbicides and green fertilizer in strawberry production. *HortScience, 45*(10), 1567–1572.

Becher, M., Talke, I. N., Krall, L., & Krämer, U. (2004). Cross-species microarray transcript profiling reveals high constitutive expression genes in shoots of the zinc hyperaccumulator *Arabidopsis halleri. The Plant Journal, 37,* 251–268.

Begonia, G. B., Miller, G. S., Begonia, M. F. T., & Burks, C. (2002). Chelate-enhanced phytoextraction of Lead-contaminated soils using coffee weed (*Sesbania exaltata* Raf.). *Bulletin of Environmental Contamination and Toxicology, 69,* 624–631.

Belouchi, A., Kwan, T., & Gros*, P. (1997). Cloning and characterization of the OsNramp family from *Oryza sativa*, a new family of membrane proteins possibly implicated in the transport of metal ions. *Plant Molecular Biology, 33,* 1085–1092.

Berken, A., Mulholland, M. M., LeDuc, D. L., & Terry, N. (2002). Genetic engineering of plants to enhance selenium phytoremediation. *Critical Reviews in Plant Sciences, 21*(6), 567–582.

Bernal, M. P., McGrath, S. P., Miller, A. J., & Baker, A. J. (1994). Comparison of the chemical changes in the rhizosphere of the nickel hyperaccumulator Alyssum murale with the non-accumulator Raphanus sativus. *Plant and Soil, 164*, 251–259.

Bernard, C., Roosens, N., Czernic, P., Lebrun, M., & Verbruggen, N. (2004). A novel CPx-ATPase from the cadmium hyperaccumulator *Thlaspi caerulescens*. *FEBS Letters, 569*(1–3), 140–148.

Berti, W. R. & Cunningham, S. D. (2000). Phytostabilization of metals. In *Phytoremediation of Toxic Metals: Using Plants to Clean up the Environment*, pp. 71–88, New York, NY: Wiley.

Bizily, S. P., Rugh, C. L., Summers, A. O., & Meagher, R. B. (1999). Phytoremediation of methylmercury pollution: merB expression in *Arabidopsis thaliana* confers resistance to organomercurials. *Proceedings of the National Academy of Sciences of the USA, 96*, 6808–6813.

Bongoua-Devisme, A. J., Akotto, O. F., Guety, T., Kouakou, S., Edith, A. A., Ndoye, F., & Diouf, D. (2019). Enhancement of phytoremediation efficiency of *Acacia mangium* using earthworms in metal contaminated soil in Bonoua, Ivory Coast. *African Journal of Biotechnology, 18*, 622–631.

Boyd, R. S. (2009). High-nickel insects and nickel hyperaccumulator plants: A review. *Insect Science, 16*(1), 19–31.

Boyd, R. S., Wall, M. A., Santos, S. R., & Davis, M. A. (2009). Variation of morphology and elemental concentrations in the California nickel hyperaccumulator *Streptanthus polygaloides* (Brassicaceae). *Northeastern Naturalist, 16*(sp5), 21–38.

Bradshaw, A. (1997). Restoration of mined lands — Using natural processes. *Ecological Engineering, 8*(4), 255–269.

Cabral, L., Soares, C. R. F. S., Giachini, A. J., & Siqueira, J. O. (2015). Arbuscular mycorrhizal fungi in phytoremediation of contaminated areas by trace elements: Mechanisms and major benefits of their applications. *World Journal of Microbiology and Biotechnology, 31*, 1655–1664.

Cai, H., Xie, P., Zeng, W., Zhai, Z., Zhou, W., & Tang, Z. (2019). Root-specific expression of rice OsHMA3 reduces shoot cadmium accumulation in transgenic tobacco. *Molecular Breeding, 39*(3), 49.

Callahan, D. L., Baker, A. J., Kolev, S. D., & Wedd, A. G. (2006). Metal ion ligands in hyperaccumulating plants. *JBIC: Journal of Biological Inorganic Chemistry, 11*, 2–12.

Chaffei, C., Pageau, K., Suzuki, A., Gouia, H., Ghorbel, M. H., & Masclaux-Daubresse, C. (2004). Cadmium toxicity induced changes in nitrogen management in Lycopersicon esculentum leading to a metabolic safeguard through an amino acid storage strategy. *Plant and Cell Physiology, 45*(11), 1681–1693.

Chang, Q., Diao, F. W., Wang, Q. F., Pan, L., Dang, Z. H., & Guo, W. (2018) Effects of arbuscular mycorrhizal symbiosis on growth, nutrient and metal uptake by maize seedlings (*Zea mays* L.) grown in soils spiked with lanthanum and cadmium. *Environmental Pollution, 241*, 607–615.

Chatterjee, J. & Chatterjee, C. (2000). Phytotoxicity of cobalt, chromium and copper in cauliflower. *Environmental Pollution, 109*(1), 69–74.

Chaturvedi, A. K., Patel, M. K., Mishra, A., Tiwari, V., & Jha, B. (2014). The SbMT-2 gene from a halophyte confers abiotic stress tolerance and modulates ROS scavenging in transgenic tobacco. *PloS One, 9*(10), e111379.

Chatzistathis, T., Alifragis, D., & Papaioannou, A. (2015). The influence of liming on soil chemical properties and on the alleviation of manganese and copper toxicity in Juglans regia, *Robinia pseudoacacia, Eucalyptus* sp. and *Populus* sp. plantations. *Journal of Environmental Management, 150*, 149–156.

Chen, S., Han, X., Fang, J., Lu, Z., Qiu, W., Liu, M., Sang, J., Jiang, J., & Zhuo, R. (2017). Sedum alfredii SaNramp6 metal transporter contributes to cadmium accumulation in transgenic *Arabidopsis thaliana. Scientific Reports, 7*(1), 13318.

Chmielowska-Bąk, J. & Deckert, J. (2021). Plant recovery after metal stress — A review. *Plants, 10*(3), 450.

Corzo Remigio, A., Chaney, R. L., Baker, A. J., Edraki, M., Erskine, P. D., Echevarria, G., & van der Ent, A. (2020). Phytoextraction of high value elements and contaminants from mining and mineral wastes: Opportunities and limitations. *Plant and Soil, 449*, 11–37.

Cunningham, S. D. & Ow, D. W. (1996). Promises and prospects of phytoremediation. *Plant Physiology, 110*(3), 715.

DalCorso, G., Fasani, E., Manara, A., Visioli, G., & Furini, A. (2019). Heavy metal pollutions: State of the art and innovation in phytoremediation. *International Journal of Molecular Sciences, 20*(14), 3412.

DalCorso, G., Manara, A., & Furini, A. (2013). An overview of heavy metal challenge in plants: From roots to shoots. *Metallomics, 5*(9), 1117–1132.

Das, N., Bhattacharya, S., & Maiti, M. K. (2016). Enhanced cadmium accumulation and tolerance in transgenic tobacco overexpressing rice metal tolerance protein gene OsMTP1 is promising for phytoremediation. *Plant Physiology and Biochemistry, 105*, 297–309.

Das, P., Datta, R., Makris, K. C., & Sarkar, D. (2010). Vetiver grass is capable of removing TNT from soil in the presence of urea. *Environmental Pollution, 158*, 1980–1983.

de Farias, V. *et al.* (2009). Phytodegradation potential of Erythrina crista-galli L., fabaceae, in petroleum-contaminated soil. *Applied Biochemistry & Biotechnology, 157*, 10–22.

Deka, R. C., Saikia, S., Biswakarma, N., Gour, N. K., & Deka, A. (2022). Nanocatalysts for exhaust emissions reduction. In Huaihe Song, Tuan Anh

Nguyen, Ghulam Yasin, Nakshatra Bahadur Singh, Ram K. Gupta (eds.). *Nanotechnology in the Automotive Industry*, pp. 511–527. Elsevier. https:// doi.org/10.1016/B978-0-323-90524-4.00024-4.

Delhaize, E., Kataoka, T., Hebb, D. M., White, R. G., & Ryan, P. R. (2003). Genes encoding proteins of the cation diffusion facilitator family that confer manganese tolerance. *The Plant Cell*, *15*(5), 1131–1142.

Dong, J., Wu, F., & Zhang, G. (2006). Influence of cadmium on antioxidant capacity and four microelement concentrations in tomato seedlings (*Lycopersicon esculentum*). *Chemosphere*, *64*(10), 1659–1666.

Dragišić Maksimović, J., Mojović, M., Maksimović, V., Römheld, V., & Nikolic, M. (2012). Silicon ameliorates manganese toxicity in cucumber by decreasing hydroxyl radical accumulation in the leaf apoplast. *Journal of Experimental Botany*, *63*(7), 2411–2420.

Dubey, S., Shri, M., Gupta, A., Rani, V., & Chakrabarty, D. (2018). Toxicity and detoxification of heavy metals during plant growth and metabolism. *Environmental Chemistry Letters*, *16*, 1169–1192.

Dufey, I., Gheysens, S., Ingabire, A., Lutts, S., & Bertin, P. (2014). Silicon application in cultivated rices (*Oryza sativa* L. and *Oryza glaberrima* Steud) alleviates iron toxicity symptoms through the reduction in iron concentration in the leaf tissue. *Journal of Agronomy and Crop Science*, *200*(2), 132–142.

Dushenkov, V., Nanda Kumar, P. B. A., Motto, H., & Raskin, I. (1995). Rhizofltration: The use of plants to remove heavy metals from aqueous streams. *Environmental Science & Technology*, *29*, 1239–1245.

Fan, M., Xiao, X., Guo, Y., Zhang, J., Wang, E., Chen, W., Lin, Y., & Wei, G. (2018). Enhanced phytoremdiation of Robinia pseudoacacia in heavy metal-contaminated soils with rhizobia and the associated bacterial community structure and function. *Chemosphere*, *197*, 729–740.

Ferretti, M., Cenni, E., Bussotti, F., & Batistoni, P. (1995). Vehicle-induced lead and cadmium contamination of roadside soil and plants in Italy. *Chemistry and Ecology*, *11*(4), 213–228.

Fourati, E., Wali, M., Vogel-Mikuš, K., Abdelly, C., & Ghnaya, T. (2016). Nickel tolerance, accumulation and subcellular distribution in the halophytes Sesuvium portulacastrum and Cakile maritima. *Plant Physiology and Biochemistry*, *108*, 295–303.

Freeman, J. L. & Salt, D. E. (2007). The metal tolerance profile of Thlaspi goesingense is mimicked in Arabidopsis thaliana heterologously expressing serine acetyl-transferase. *BMC Plant Biology*, *7*, 1–10.

Freeman, J. L., Garcia, D., Kim, D., Hopf, A., & Salt, D. E. (2005). Constitutively elevated salicylic acid signals glutathione-mediated nickel tolerance in Thlaspi nickel hyperaccumulators. *Plant Physiology*, *137*(3), 1082–1091.

Freeman, R. E., Wicks, A. C., & Parmar, B. (2004). Stakeholder theory and "the corporate objective revisited". *Organization Science, 15*(3), 364–369.

Galeas, M. L., Zhang, L. H., Freeman, J. L., Wegner, M., & Pilon-Smits, E. A. (2007). Seasonal fluctuations of selenium and sulfur accumulation in selenium hyperaccumulators and related nonaccumulators. *The New Phytologist, 173*(3), 517–525.

Gangwar, S. & Singh, V. P. (2011). Indole acetic acid differently changes growth and nitrogen metabolism in *Pisum sativum* L. seedlings under chromium (VI) phytotoxicity: Implication of oxidative stress. *Scientia Herticulturae, 129*(2), 321–328.

Ghazaryan, K. A., Movsesyan, H. S., Minkina, T. M., Sushkova, S. N., & Rajput, V. D. (2019). The identification of phytoextraction potential of Melilotus officinalis and Amaranthus retroflexus growing on copper-and molybdenum-polluted soils. *Environmental Geochemistry and Health, 43*, 1327–1335.

Ghosh, M. & Singh, S. P. (2005). A review on phytoremediation of heavy metals and utilization of it's by products. *Asia Pacific Journal of Energy and Environment, 6*(4), 18.

Gong, Z., Xiong, L., Shi, H., Yang, S., Herrera-Estrella, L. R., Xu, G., Chao, D. Y., Li, J., Wang, P. Y., Qin, F., & Zhu, J. K. (2020). Plant abiotic stress response and nutrient use efficiency. *Science China Life Sciences, 63*, 635–674.

Guo, J. & Chi, J. (2014). Effect of Cd-tolerant plant growth-promoting rhizobium on plant growth and Cd uptake by Lolium multiflorum Lam. and *Glycine max* (L.) Merr. in Cd-contaminated soil. *Plant and Soil, 375*, 205–214.

Gupta, D. K. (2002). *Ecophysiological Studies on Cicer arietinum L. Growing under Fly-Ash Stress Condition.* PhD Thesis, Lucknow University, India.

Gupta, D. K. (2013). *Plant Based Remediation Process.* Springer.

Gupta, D. K. & Chatterjee, S. (2014). *Heavy Metal Remediation: Transport and Accumulation in Plants.* New York, NY: Nova Science Publishers.

Gupta, D. K. & Chatterjee, S. (2017). *Arsenic Contamination in the Environment: The Issues and Solutions.* Springer.

Gupta, D. K., Corpas, F.J., & Palma, J. M. (2013) *Heavy Metal Stress in Plants.* Springer.

Hajiboland, R., Bahrami Rad, S., Barceló, J., & Poschenrieder, C. (2013). Mechanisms of aluminum-induced growth stimulation in tea (*Camellia sinensis*). *Journal of Plant Nutrition and Soil Science, 176*(4), 616–625.

Hannink, N., Subramanian, N., Rosser, S. J., Basran, A., Murray, J. A. H., & Shanks, J. V. (2007). Enhanced transformation of TNT by tobacco plants expressing a bacterial nitroreductase. *International Journal of Phytoremediation, 9*, 385–401.

Harumain, Z. A., Parker, H. L., Muñoz García, A., Austin, M. J., McElroy, C. R., Hunt, A. J., James, H. C., John A. M., Christopher, W. N., Anderson, Luca, C, Graedel, T. E., & Rylott, E. L. (2017). Toward financially viable phytoextraction and production of plant-based palladium catalysts. *Environmental Science & Technology*, *51*(5), 2992–3000.

Hasan, M. M., Uddin, M. N., Ara-Sharmeen, I. F., Alharby, H., Alzahrani, Y., Hakeem, K. R., & Zhang, L. (2019). Assisting phytoremediation of heavy metals using chemical amendments. *Plants*, *8*, 295. https://doi.org/10.3390/plants8090295.

Helaoui, S., Mkhinini, M., Boughattas, I., Bousserrhine, N., & Banni, M. (2023). Nickel toxicity and tolerance in plants. In Mohammad Anwar Hossain, AKM Zakir Hossain, Sylvain Bourgerie, Masayuki Fujita, Om Parkash Dhankher, Parvez Haris (eds.), *Heavy Metal Toxicity and Tolerance in Plants: A Biological, Omics, and Genetic Engineering Approach*, pp. 231–250. John Wiley & Sons Ltd.

Hernández, L. E., Garate, A., & Carpena-Ruiz, R. (1997). Effects of cadmium on the uptake, distribution and assimilation of nitrate in *Pisum sativum*. *Plant and Soil*, *189*, 97–106.

Hernández, L. E., Ortega-Villasante, C., Montero-Palmero, M. B., Escobar, C., & Carpena, R. O. (2012). Heavy metal perception in a microscale environment: A model system using high doses of pollutants. In D. Gupta & L. Sandalio (eds.), *Metal Toxicity in Plants: Perception, Signaling and Remediation*, pp. 23–39, Berlin, Heidelberg: Springer.

Hoehne, L., de Lima, C. V., Martini, M. C., Altmayer, T., Brietzke, D. T., Finatto, J., *et al.* (2016) Addition of vermicompost to heavy metal-contaminated soil increases the ability of black oat (*Avena strigosa* Schreb) plants to remove Cd, Cr, and Pb. *Water, Air, Soil & Pollution*, *227*, 443.

Hossain, M. A., Hossain, A. Z., Bourgerie, S., Fujita, M., Dhankher, O. P., & Haris, P. (eds.) (2023). *Heavy Metal Toxicity and Tolerance in Plants: A Biological, Omics, and Genetic Engineering Approach*. John Wiley & Sons.

Hossain, M. A., Piyatida, P., da Silva, J. A. T., & Fujita, M. (2012). Molecular mechanism of heavy metal toxicity and tolerance in plants: Central role of glutathione in detoxification of reactive oxygen species and methylglyoxal and in heavy metal chelation. *Journal of Botany*, *2012*, 37.

Huang, X., Jain, P. K., El-Sayed, I. H., & El-Sayed, M. A. (2007). Gold nanoparticles: Interesting optical properties and recent applications in cancer diagnostics and therapy. *Nanomedicine (London)*, *2*(5), 681–693. https://doi.org/10.2217/17435889.2.5.681.

Hui, C. Y., Ma, B. C., Hu, S. Y., & Wu, C. (2023). Tailored bacteria tackling with environmental mercury: Inspired by natural mercuric detoxification operons. *Environmental Pollution*, *341*, 123016.

Hunt, A. J., Farmer, T. J., & Clark, J. H. (2013). Elemental sustainability and the importance of scarce element recovery. In A. Hunt (ed.), *Element Recovery and Sustainability*. The Royal Society.

Hunt, A. J., Anderson, C. W., Bruce, N., García, A. M., Graedel, T. E., Hodson, M., Meech, J. A., Nassar, N. T., Parker, H. L., Rylott, E. L., & Clark, J. H. (2014). Phytoextraction as a tool for green chemistry. *Green Processing and Synthesis, 3*(1), 3–22.

Husaini, Y. & Rai, L. C. (1991). Studies on nitrogen and phosphorus metabolism and the photosynthetic electron transport system of *Nostoc linckia* under cadmium stress. *Journal of Plant Physiology, 138*(4), 429–435.

Imran, M., Mahmood, A., Römheld, V., & Neumann, G. (2013). Nutrient seed priming improves seedling development of maize exposed to low root zone temperatures during early growth. *European Journal of Agronomy, 49*, 141–148.

Jacob, J. M., Karthik, C., Saratale, R. G., Kumar, S. S., Prabakar, D., Kadirvelu, K., & Pugazhendhi, A. (2018). Biological approaches to tackle heavy metal pollution: A survey of literature. *Journal of Environmental Management, 217*, 56–70.

Jacobs, A., Drouet, T., & Noret, N. (2018). Field evaluation of cultural cycles for improved cadmium and zinc phytoextraction with *Noccaea caerulescens*. *Plant Soil, 430*, 381–394.

Jarup, L. (2003). Hazards of heavy metal contamination. *British Medical Bulletin, 68*(1), 167–182.

Ji, P., Tang, X., Jiang, Y., *et al.* (2015). Potential of gibberellic acid 3 (GA3) for enhancing the phytoremediation efficiency of *Solanum nigrum* L. *Bulletin of Environmental Contamination and Toxicology, 95*, 810–814.

Johar, P., McElroy, C. R., Rylott, E. L., Matharu, A. S., & Clark, J. H. (2022). Biologically bound nickel as a sustainable catalyst for the selective hydrogenation of cinnamaldehyde. *Applied Catalysis B: Environmental, 306*, 121105.

Just, C. L. & Schnoor, J. L. (2004). Phytophotolysis of hexahydro-1,3,5- trinitro-1,3,5-triazine (RDX) in leaves of reed canary. *Environmental Science & Technology, 38*, 290–295.

Kaur, P., Bali, S., Sharma, A., Vig, A. P., & Bhardwaj, R. (2018). Role of earthworms in phytoremediation of cadmium (Cd) by modulating the antioxidative potential of *Brassica juncea* L. *Applied Soil Ecology, 124*, 306–316.

Kaya, C., Ashraf, M., Sonmez, O., Aydemir, S., Tuna, A. L., & Cullu, M. A. (2009). The influence of arbuscular mycorrhizal colonisation on key growth parameters and fruit yield of pepper plants grown at high salinity. *Scientia Horticulturae, 121*(1), 1–6.

Kenderešová, L., Staňová, A., Pavlovkin, J., Ďurišová, E., Nadubinská, M., Čiamporová, M., & Ovečka, M. (2012). Early Zn2+-induced effects on membrane potential account for primary heavy metal susceptibility in tolerant and sensitive Arabidopsis species. *Annals of Botany, 110*(2), 445–459.

Keunen, E., Remans, T., Bohler, S., Vangronsveld, J., & Cuypers, A. (2011). Metal-induced oxidative stress and plant mitochondria. *International Journal of Molecular Sciences, 12*(10), 6894–6918.

Khalid, A., Farid, M., Zubair, M., Rizwan, M., Iftikhar, U., Ishaq, H. K., Farid, S., Latif, U., Hina, K., & Ali, S. (2020). Efcacy of Alternanthera bettzickiana to remediate copper and cobalt contaminated soil physiological and biochemical alterations. *International Journal of Environmental Research, 14*, 243–255.

Khalid, S., Shahid, M., Niazi, N. K., Murtaza, B., Bibi, I., & Dumat, C. (2017). A comparison of technologies for remediation of heavy metal contaminated soils. *Journal of Geochemical Exploration, 182*, 247–268.

Kirk, J., Klironomos, J., Lee, H., & Trevors, J. T. (2005). The effects of perennial ryegrass and alfalfa on microbial abundance and diversity in petroleum contaminated soil. *Environmental Pollution, 133*, 455–465.

Kodituwakku, K. A. R. K. & Yatawara, M. (2020). Phytoremediation of industrial sewage sludge with *Eichhornia crassipes, Salvinia molesta* and *Pistia stratiotes* in batch fed free water flow constructed wetlands. *Bulletin of Environmental Contamination and Toxicology, 104*, 627–633.

Kraemer, S. M., Kyes, S. A., Aggarwal, G., Springer, A. L., Nelson, S. O., Christodoulou, Z., Wang, W., Levin, E., Newbold, C. I., Myler, P. J., & Smith, J. D. (2007). Patterns of gene recombination shape var gene repertoires in Plasmodium falciparum: Comparisons of geographically diverse isolates. *BMC Genomics, 8*, 1–18.

Krämer, U. (2010). Metal hyperaccumulation in plants. *Annual Review of Plant Biology, 61*, 517–534.

Krishnamurti, G. S. R., Cieslinski, G., Huang, P. M., & Van Rees, K. C. J. (1997). *Kinetics of Cadmium Release from Soils as Influenced by Organic Acids: Implication in Cadmium Availability*, Vol. 26, No. 1, pp. 271–277. American Society of Agronomy, Crop Science Society of America, and Soil Science Society of America.

Krishnasamy, S., Lakshmanan, R., & Ravichandran, M. (2022). Phytoremediatiation of metal and metalloid pollutants from farmland: An *in-situ* soil conservation. In *Biodegradation Technology of Organic and Inorganic Pollutants.* IntechOpen. https://doi.org/10.5772/intechopen.98659.

Kukreja, S. & Goutam, U. (2012). Phytoremediation: A new hope for the environment. *Frontiers on Recent Developments in Plant Science, 1*, 149–171.

Kumar, S., Prasad, S., Yadav, K. K., Shrivastava, M., Gupta, N., Nagar, S., Bach, Q. V., Kamyab, H., Khana, S. S., & Yadav, S. (2019). Hazardous heavy metals contamination of vegetables and food chain: Role of sustainable remediation approaches — A review. *Environmental Research, 179*, 108792.

Kumchai, J., Huang, J. Z., Lee, C. Y., Chen, F. C., & Chin, S. W. (2013). Proline partially overcomes excess molybdenum toxicity in cabbage seedlings grown in vitro. *Genetics and Molecular Research, 12,* 5589–5601.

Kuriakose, S. V. & Prasad, M. N. V. (2008). Cadmium stress affects seed germination and seedling growth in *Sorghum bicolor* (L.) Moench by changing the activities of hydrolyzing enzymes. *Plant Growth Regulation, 54,* 143–156.

Lasat, M. M. (2002). Phytoextraction of toxic metals: A review of biological mechanisms. *Journal of Environmental Quality, 31*(1), 109–120.

Lea, P. J. & Miflin, B. J. (2003). Glutamate synthase and the synthesis of glutamate in plants. *Plant Physiology and Biochemistry, 41*(6–7), 555–564.

Lee, J. D., Bilyeu, K. D., & Shannon, J. G. (2007). Genetics and breeding for modified fatty acid profile in soybean seed oil. *Journal of Crop Science and Biotechnology, 10*(4), 201–210.

Leng, Y., Li, Y., Ma, Y. H., *et al.* (2020). Abscisic acid modulates differential physiological and biochemical responses of roots, stems, and leaves in mung bean seedlings to cadmium stress. *Environmental Science and Pollution Research, 28*(5), 6030–6043.

Li, L., Huang, X., Borthakur, D., & Ni, H. (2012). Photosynthetic activity and antioxidative response of seagrass *Thalassia hemprichii* to trace metal stress. *Acta Oceanologica Sinica, 31,* 98–108.

Li, S., Xu, Z., Cheng, X., & Zhang, Q. (2008). Dissolved trace elements and heavy metals in the Danjiangkou Reservoir, China. *Environmental Geology, 55,* 977–983.

Li, Y.-M., Chaney, R. L., Brewer, E., Roseberg, R., Angle, J. S., Baker, A. J. M., Reeves, R. D., & Nelkin, J. (2003). Development of a technology for commercial phytoextraction of nickel: Economic and technical considerations. *Plant & Soil, 249,* 107–115.

Liang, Y., Sun, W., Zhu, Y. G., & Christie, P. (2007). Mechanisms of silicon-mediated alleviation of abiotic stresses in higher plants: A review. *Environmental Pollution, 147*(2), 422–428.

Liu, D., Jiang, W., & Li, M. (1992). Effects of trivalent and hexavalent chromium on root growth and cell division of Allium cepa. *Hereditas, 117*(1), 23–29.

Liu, S., Yang, B., Liang, Y., Xiao, Y., & Fang, J. (2020a). Prospect of phytoremediation combined with other approaches for remediation of heavy metal-polluted soils. *Environmental Science and Pollution Research, 27,* 16069–16085.

Liu, W. J., Tian, K., Jiang, H., Zhang, X. S., Ding, H. S., & Yu, H. Q. (2012). Selectively improving the bio-oil quality by catalytic fast pyrolysis of heavy-metal-polluted biomass: Take copper (Cu) as an example. *Environmental Science & Technology, 46*(14), 7849–7856.

Liu, Y., Kang, T., Cheng, J. S., Yi, Y. J., Han, J. J., Cheng, H. L., Li, Q., Tang, N., & Liang, M. X. (2020b). Heterologous expression of the metallothionein Ppmt2 gene from Physcomitrella patens confers enhanced tolerance to heavy metal stress on transgenic Arabidopsis plants. *Plant Growth Regulation, 90*, 63–72.

Lopez, A. M. Q., Silva, A. L. D. S., Maranhão, F. C. D. A., & Ferreira, L. F. R. (2022). Plant growth promoting bacteria: Aspects in metal bioremediation and phytopathogen management. In *Microbial Biocontrol: Sustainable Agriculture and Phytopathogen Management,* Vol. 1, pp. 51–78. Springer International Publishing, Cham.

Losfeld, G., de La Blache, P. V., Escande, V., & Grison, C. (2012). Zinc hyperaccumulating plants as renewable resources for the chlorination process of alcohols. *Green Chemistry Letters and Reviews, 5*(3), 451–456.

Lu, Y. P., Li, Z. S., Drozdowicz, Y. M., Hörtensteiner, S., Martinoia, E., & Rea, P. A. (1998). AtMRP2, an Arabidopsis ATP binding cassette transporter able to transport glutathione S-conjugates and chlorophyll catabolites: Functional comparisons with AtMRP1. *The Plant Cell, 10*(2), 267–282.

Ma, J. F., Goto, S., Tamai, K., & Ichii, M. (2001). Role of root hairs and lateral roots in silicon uptake by rice. *Plant Physiology, 127*(4), 1773–1780.

Macek, T., Macková, M., & Káš, J. (2000). Exploitation of plants for the removal of organics in environmental remediation. *Biotechnology Advances, 18*(1), 23–34.

Mahmoud, S. A., Mohamed, B. S., & Killa, H. M. (2021). Synthesis of different sizes TiO_2 and photovoltaic performance in dye-sensitized solar cells. *Frontiers in Materials, 8*, 714835.

Maleva, M. G., Nekrasova, G. F., Borisova, G. G., Chukina, N. V., & Ushakova, O. S. (2012). Effect of heavy metals on photosynthetic apparatus and antioxidant status of Elodea. *Russian Journal of Plant Physiology, 59*, 190–197.

Mani, D. & Kumar, C. (2014). Biotechnological advances in bioremediation of heavy metals contaminated ecosystems: An overview with special reference to phytoremediation. *International Journal of Environmental Science and Technology, 11*, 843–872.

Manzoor, M., Gul, I., Manzoor, A., Kamboh, U. R., Hina, K., Kallerhof, J., & Arshad, M. (2020). Lead availability and phytoextraction in the rhizosphere of Pelargonium species. *Environmental Science and Pollution Research, 27*, 39753–39762. https://doi.org/10.1007/s11356-020-08226-0.

Mäser, P., Thomine, S., Schroeder, J. I., Ward, J. M., Hirschi, K., Sze, H., Talke, I. N., Amtmann, A., Maathuis, F. J., Sanders, D., Harper, J. F., & Guerinot, M. L. (2001). Phylogenetic relationships within cation transporter families of Arabidopsis. *Plant Physiology, 126*(4), 1646–1667.

Mataruga, Z., Jarić, S., Marković, M., Pavlović, M., Pavlović, D., Jakovljević, K., Mitrović, M., & Pavlović, P. (2020) Evaluation of Salix alba, Juglans regia and Populus nigra as biomonitors of PTEs in the riparian soils of the Sava River. *Environmental Monitoring and Assessment, 192*, 131.

McGrath, S. P., Zhao, F. J., & Lombi, E. (2002). Phytoremediation of metals, metalloids, and radionuclides. *Advances in Agronomy, 75*, 1–56.

Mendoza-Cózatl, D. G., Jobe, T. O., Hauser, F., & Schroeder, J. I. (2011). Long-distance transport, vacuolar sequestration, tolerance, and transcriptional responses induced by cadmium and arsenic. *Current Opinion in Plant Biology, 14*(5), 554–562.

Mendoza-Hernández, J. C., Vázquez-Delgado, O. R., Castillo-Morales, M., Varela-Caselis, J. L., Santamaría-Juárez, J. D., Olivares-Xometl, O., Morales, J. A., & Pérez-Osorio, G. (2019). Phytoremediation of mine tailings by Brassica juncea inoculated with plant growth-promoting bacteria. *Microbiological Research, 228*, 126308.

Mesjasz-Przybyłowicz, J., Nakonieczny, M, Migula, P, Augustyniak, M., Tarnawska, M., Reimold, W. U., Koeberl, C., Przybyłowicz, W., & Głowacka, E. (2004). Uptake of cadmium, lead nickel and zinc from soil and water solutions by the nickel hyperaccumulator Berkheya coddii. *Acta Biologica Cracoviensia Series Botanica, 46*, 75–85.

Monaci, F., Trigueros, D., Mingorance, M. D., & Rossini-Oliva, S. (2019). Phytostabilization potential of *Erica australis* L. and *Nerium oleander* L.: A comparative study in the Riotinto mining area (SW Spain). *Environmental Geochemistry and Health, 42*, 2345–2360.

Montanini, B., Blaudez, D., Jeandroz, S., Sanders, D., & Chalot, M. (2007). Phylogenetic and functional analysis of the Cation Diffusion Facilitator (CDF) family: Improved signature and prediction of substrate specificity. *BMC Genomics, 8*(1), 1–16.

Moreno, F. N., Anderson, C. W. N., Stewart, R. B., & Robinson, B. H. (2008). Phytofltration of mercury-contaminated water: Volatilization and plant-accumulation aspects. *Environmental and Experimental Botany, 62*, 78–85.

Mura, P., Brunet, B., Papet, Y., & Hauet, T. (2004). Cannabis sativa var. indica: une plante complexe aux effets pervers. In *Annales de toxicologie analytique* Vol. 16, No. 1, pp. 7–17. EDP Sciences.

Nagajyoti, P. C., Lee, K. D., & Sreekanth, T. V. M. (2010). Heavy metals, occurrence and toxicity for plants: A review. *Environmental Chemistry Letters, 8*, 199–216.

Nedjimi, B. (2009). Can calcium protect *Atriplex halimus* subsp. schweinfurthii against cadmium toxicity? *Acta Botanica Gallica, 156*(3), 391–397.

Nedjimi, B. (2020). Germination characteristics of *Peganum harmala* L. (Nitrariaceae) subjected to heavy metals: Implications for the use in polluted dryland restoration. *International Journal of Environmental Science and Technology, 17*(4), 2113–2122.

Nedjimi, B. (2021). Phytoremediation: A sustainable environmental technology for heavy metals decontamination. *SN Applied Sciences, 3*(3), 286.

Nedjimi, B. & Daoud, Y. (2009a). Ameliorative effect of $CaCl_2$ on growth, membrane permeability and nutrient uptake in Atriplex halimus subsp. schweinfurthii grown at high (NaCl) salinity. *Desalination, 249*(1), 163–166.

Nedjimi, B. & Daoud, Y. (2009b). Cadmium accumulation in Atriplex halimus subsp. schweinfurthii and its influence on growth, proline, root hydraulic conductivity and nutrient uptake. *Flora, 204,* 316–324.

Nkrumah, P. N., Echevarria, G., Erskine, P. D., & van der Ent, A. (2018). Phytomining: Using plants to extract valuable metals from mineralised wastes. In M. J. Clifford, R. K. Perrons, S. H. Ali, & T. A. Grice (eds.), *Extracting Innovations: Mining, Energy, and Technological Change in the Digital Age Taylor.* https://doi.org/10.1201/b22353-23

Ortiz, D. F., Kreppel, L., Speiser, D. M., Scheel, G., McDonald, G., & Ow, D. W. (1992). Heavy metal tolerance in the fission yeast requires an ATP-binding cassette-type vacuolar membrane transporter. *The EMBO Journal, 11*(10), 3491–3499.

Ortiz, D. F., Ruscitti, T., McCue, K. F., & Ow, D. W. (1995). Transport of metal-binding peptides by HMT1, a fission yeast ABC-type vacuolar membrane protein (∗). *Journal of Biological Chemistry, 270*(9), 4721–4728.

Oustriere, N., Marchand, L., Roulet, E., & Mench, M. (2017). Rhizofiltration of a Bordeaux mixture effluent in pilot-scale constructed wetland using *Arundo donax* L. coupled with potential Cu-ecocatalyst production. *Ecological Engineering, 105,* 296–305.

Papoyan, A. & Kochian, L. V. (2004). Identification of Thlaspi caerulescens genes that may be involved in heavy metal hyperaccumulation and tolerance. Characterization of a novel heavy metal transporting ATPase. *Plant Physiology, 136*(3), 3814–3823.

Parker, H. L., Rylott, E. L., Hunt, A. J., Dodson, J. R., Taylor, A. F., Bruce, N. C., & Clark, J. H. (2014). Supported palladium nanoparticles synthesized by living plants as a catalyst for Suzuki-Miyaura reactions. *Plos One, 9*(1), e87192.

Patil, U. P. & Patil, S. S. (2021). Natural feedstock in catalysis: A sustainable route towards organic transformations. *Topics in Current Chemistry, 379*(5), 36.

Peer, W. A., Baxter, I. R., Richards, E. L., Freeman, J. L., & Murphy, A. S. (2005). Phytoremediation and hyperaccumulator plants. In Tamas (ed.), *Molecular Biology of Metal Homeostasis and Detoxification. Topics in Current Genetics,* Vol. 14. Springer, Berlin, Heidelberg.

Pena, L. B., Barcia, R. A., Azpilicueta, C. E., Méndez, A. A., & Gallego, S. M. (2012). Oxidative post translational modifications of proteins related to cell cycle are involved in cadmium toxicity in wheat seedlings. *Plant Science, 196,* 1–7.

Peng, K., Li, X., Luo, C., & Shen, Z. (2006). Vegetation composition and heavy metal uptake by wild plants at three contaminated sites in Xiangxi area, China. *Journal of Environmental Science and Health, Part A, 41*(1), 65–76.

Peto, A., Lehotai, N., Lozano-Juste, J., León, J., Tari, I., Erdei, L., & Kolbert, Z. (2011). Involvement of nitric oxide and auxin in signal transduction of copper-induced morphological responses in Arabidopsis seedlings. *Annals of Botany, 108*(3), 449–457.

Pilon-Smits, E. (2005). Phytoremediation. *Annual Review of Plant Biology, 56,* 15–39.

Poirier Y. & Bucher M. (2002). Phosphate transport and homeostasis in Arabidopsis. In *Arabidopsis Book*, Vol. 1, e0024. https://doi.org/10.1199/tab.0024.

Prapagdee, B. & Khonsue, N. (2015). Bacterial-assisted cadmium phytoremediation by Ocimum gratissimum L. in polluted agricultural soil: A field trial experiment. *International Journal of Environmental Science and Technology, 12,* 3843–3852.

Prasad, T. N. V. K. V., Sudhakar, P., Sreenivasulu, Y., Latha, P., Munaswamy, V., Reddy, K. R., Sreeprasad, T. S., Sajanlal, P. R., & Pradeep, T. (2012). Effect of nanoscale zinc oxide particles on the germination, growth and yield of peanut. *Journal of Plant Nutrition, 35*(6), 905–927.

Rai, P. K., Kim, K. H., Lee, S. S., & Lee, J. H. (2020). Molecular mechanisms in phytoremediation of environmental contaminants and prospects of engineered transgenic plants/microbes. *Science of the Total Environment, 705,* 135858.

Ramakrishna, B. & Rao, S. S. R. (2012). 24-Epibrassinolide alleviated zinc-induced oxidative stress in radish (*Raphanus sativus* L.) seedlings by enhancing antioxidative system. *Plant Growth Regulation, 68,* 249–259.

Rascio, N. & Navari-Izzo, F. (2011). Heavy metal hyperaccumulating plants: How and why do they do it? And what makes them so interesting? *Plant Science, 180*(2), 169–181.

Raza, A., Habib, M., Kakavand, S. N., Zahid, Z., Zahra, N., Sharif, R., & Hasanuzzaman, M. (2020). Phytoremediation of cadmium: Physiological, biochemical, and molecular mechanisms. *Biology, 9*(7), 177.

Reeves, R. D. & Baker, A. J. M. (2000). Metal accumulating plants. In: Raskin, I. and Finsley, B.D. (eds.), *Phytoremediation of Toxic Metals: Using Plants to Clean up the Environment*, pp. 193–229. Wiley, New York, NY.

Reeves, R. D. & Brooks, R. R. (1983). Hyperaccumulation of mining areas of Central Europe. *Environmental Pollution Series A, Ecological and Biological, 31*(4), 277–285.

Reeves, R. D., van der Ent, A., Echevarria, G., Isnard, S., & Baker, A. J. M. (2021). Global distribution and ecology of hyperaccumulator plants. In van der Ent, A., Baker, A. J., Echevarria, G., Simonnot, MO., Morel, J. L. (eds.),

Agromining: Farming for Metals. Mineral Resource Reviews. Springer, Cham. https://doi.org/10.1007/978-3-030-58904-2_7.

Rehman, M., Maqbool, Z., Peng, D., & Liu, L. (2019). Morpho-physiological traits, antioxidant capacity and phytoextraction of copper by ramie (*Boehmeria nivea* L.) grown as fodder in copper-contaminated soil. *Environmental Science and Pollution Research, 26*, 5851–5861.

Ricachenevsky, F. K., Menguer, P. K., Sperotto, R. A., Williams, L. E., & Fett, J. P. (2013). Roles of plant metal tolerance proteins (MTP) in metal storage and potential use in biofortification strategies. *Frontiers in Plant Science, 4*, 144.

Robinson, B. H., Brooks, R. R., Howes, A. W., Kirkman, J. H., & Gregg, P. E. H. (1997). The potential of the high-biomass nickel hyperaccumulator *Berkheya coddii* for phytoremediation and phytomining. *Journal of Geochemical Exploration, 60*(2), 115–126.

Rogalla, H. & Römheld, V. (2002). Role of leaf apoplast in silicon-mediated manganese tolerance of *Cucumis sativus* L. *Plant, Cell & Environment, 25*(4), 549–555.

Rugh, C. L., Senecof, J. F., Meagher, R. B., & Merkle, S. A. (1998). Development of transgenic yellow poplar for mercury phytoremediation. *Nature Biotechnology, 16*, 925–928.

Salt, D. E., Prince, R. C., Baker, A. J., Raskin, I., & Pickering, I. J. (1999). Zinc ligands in the metal hyperaccumulator *Thlaspi caerulescens* as determined using X-ray absorption spectroscopy. *Environmental Science & Technology, 33*(5), 713–717.

Sampaio, C. J. S., de Souza, J. R. B., Damião, A. O., Bahiense, T. C., & Roque, M. R. A. (2019). Biodegradation of polycyclic aromatic hydrocarbons (PAHs) in a diesel oil-contaminated mangrove by plant growth-promoting rhizobacteria. *3 Biotech, 9*, 155. https://doi.org/10.1007/s13205-019-1686-8.

Santana, N. A., Ferreira, P. A. A., Tarouco, C. P., Schardong, I. S., Antoniolli, Z. I., Nicoloso, F. T., & Jacques, R. J. S. (2019). Earthworms and mycorrhization increase copper phytoextraction by *Canavalia ensiformis* in sandy soil. *Ecotoxicology and Environmental Safety, 182*, 109383.

Santos-Francés, F., Martínez-Graña, A., Alonso Rojo, P., & García Sánchez, A. (2017). Geochemical background and baseline values determination and spatial distribution of heavy metal pollution in soils of the Andes mountain range (Cajamarca-Huancavelica, Peru). *International Journal of Environmental Research and Public Health, 14*(8), 859.

Sarret, G., Saumitou-Laprade, P., Bert, V., Proux, O., Hazemann, J. L., Traverse, A., Marcus, M. A., & Manceau, A. (2002). Forms of zinc accumulated in the hyperaccumulator *Arabidopsis halleri. Plant Physiology, 130*(4), 1815–1826.

Schäfer, C., Simper, H., & Hofmann, B. (1992). Glucose feeding results in coordinated changes of chlorophyll content, ribulose-1, 5-bisphosphate carboxylase-oxygenase activity and photosynthetic potential in photoautrophic suspension cultured cells of Chenopodium rubrum. *Plant, Cell & Environment, 15*(3), 343–350.

Schnoor, J. L., Light, L. A., McCutcheon, S. C., Wolfe, N. L., & Carreia, L. H. (1995). Phytoremediation of organic and nutrient contaminants. *Environmental Science & Technology, 29*(7), 318A–323A.

Severne, B. C. & Brooks, R. R. (1972). A nickel-accumulating plant from Western Australia. *Planta, 103*, 91–94.

Shabani, L., Sabzalian, M. R., & Pour, S. M. (2016). Arbuscular mycorrhiza affects nickel translocation and expression of ABC transporter and metallothionein genes in *Festuca arundinacea*. *Mycorrhiza, 26*, 67–76.

Shah, K. & Nongkynrih, J. M. (2007). Metal hyperaccumulation and bioremediation. *Biologia Plantarum, 51*, 618–634.

Shah, V. & Daverey, A. (2020) Phytoremediation: A multidisciplinary approach to clean up heavy metal contaminated soil. *Environmental Technology & Innovation, 18*, 100774.

Shahzad, T., Chenu, C., Genet, P., Barot, S., Perveen, N., Mougin, C., & Fontaine, S. (2015). Contribution of exudates, arbuscular mycorrhizal fungi and litter depositions to the rhizosphere priming effect induced by grassland species. *Soil Biochemistry, 80*, 146–155.

Sharma, A. and Nagpal, A. K. (2020). Contamination of vegetables with heavy metals across the globe: Hampering food security goal. *Journal of Food Science & Technology, 57*, 391–403.

Sharma, N. C., Sahi, S. V., Nath, S., Parsons, J. G., Gardea-Torresde, J. L., & Pal, T. (2007). Synthesis of plant-mediated gold nanoparticles and catalytic role of biomatrix-embedded nanomaterials. *Environmental Science & Technology, 41*(14), 5137–5142.

Sharma, P., Bakshi, P., Kour, J., Singh, A. D., Dhiman, S., Kumar, P., Sharma I. A., Mir, B. A., & Bhardwaj, R. (2020) PGPR and earthworm-assisted phytoremediation of heavy metals. In S. A. Bhat *et al.* (eds.), *Earthworm Assisted Remediation of Effluents and Wastes*, pp. 227–245. Springer, Singapore.

Sharma, S. S. & Dietz, K. J. (2009). The relationship between metal toxicity and cellular redox imbalance. *Trends in Plant Science, 14*(1), 43–50.

Siddiqui, M. H., Al-Whaibi, M. H., Ali, H. M., Sakran, A. M., Basalah, M. O., & AlKhaishany, M. Y. (2013). Mitigation of nickel stress by the exogenous application of salicylic acid and nitric oxide in wheat. *Australian Journal of Crop Science, 7*(11), 1780–1788.

Simon, S., Sibuyi, N. R. S., Fadaka, A. O., Meyer, S., Josephs, J., Onani, M. O., Meyer, M., & Madiehe, A. M. (2022). Biomedical applications of plant extract-synthesized silver nanoparticles. *Biomedicines, 10*(11), 2792.

Singh, O. V., Labana, S., Pandey, G., Budhiraja, R., & Jain, R. K. (2003). Phytoremediation: An overview of metallic ion decontamination from soil. *Applied Microbiology and Biotechnology, 61*, 405–412.

Singh, S., Parihar, P., Singh, R., Singh, V. P., & Prasad, S. M. (2016). Heavy metal tolerance in plants: Role of transcriptomics, proteomics, metabolomics, and ionomics. *Frontiers in Plant Science, 6*, 1143.

Sirhindi, G., Mir, M. A., Sharma, P., Gill, S. S., Kaur, H., & Mushtaq, R. (2015). Modulatory role of jasmonic acid on photosynthetic pigments, antioxidants and stress markers of *Glycine max* L. under nickel stress. *Physiology and Molecular Biology of Plants, 21*, 559–565.

Song, C., Yan, Y., Rosado, A. R., Zhang, Z., & Castellarin, S. D. (2019). ABA alleviates uptake and accumulation of zinc in grapevine (*Vitis vinifera* L.) by inducing expression of ZIP and detoxification-related genes. *Frontiers in Plant Science, 10*, 872.

Srivastava, S., Srivastava, A. K., Suprasanna, P., & D'souza, S. F. (2013). Identification and profiling of arsenic stress-induced microRNAs in *Brassica juncea*. *Journal of Experimental Botany, 64*(1), 303–315.

Striker, G. G. (2011). Time is on our side: The importance of considering a recovery periods when assessing flooding tolerance in plants. *Ecological Research, 27*, 983–987.

Suman, J., Uhlik, O., Viktorova, J., & Macek, T. (2018). Phytoextraction of heavy metals: A promising tool for clean-up of polluted environment? *Frontiers in Plant Science, 9*, 1476.

Sundaramoorthy, S., Goh, J. B. G., Rafee, S., & Murata-Hori, M. (2010). Mitotic Golgi vesiculation involves mechanisms independent of Ser25 phosphorylation of GM130. *Cell Cycle, 9*(15), 3172–3177.

Susarla, S., Medina, V. F., & McCutcheon, S. C. (2002). Phytoremediation: An ecological solution to organic chemical contamination. *Ecological Engineering, 18*, 647–658.

Szuba, A., Marczak, Ł., & Kozłowski, R. (2020). Role of the proteome in providing phenotypic stability in control and ectomycorrhizal poplar plants exposed to chronic mild Pb stress. *Environmental Pollution, 264*, 114585.

Talke, I. N., Hanikenne, M., & Krämer, U. (2006). Zinc-dependent global transcriptional control, transcriptional deregulation, and higher gene copy number for genes in metal homeostasis of the hyperaccumulator *Arabidopsis halleri*. *Plant Physiology, 142*(1), 148–167.

Tamás, L., Mistrík, I., Huttová, J., Halušková, L. U., Valentovičová, K., & Zelinová, V. (2010). Role of reactive oxygen species-generating enzymes and hydrogen peroxide during cadmium, mercury and osmotic stresses in barley root tip. *Planta, 231*, 221–231.

Tamás, M. J., Sharma, S. K., Ibstedt, S., Jacobson, T., & Christen, P. (2014). Heavy metals and metalloids as a cause for protein misfolding and aggregation. *Biomolecules, 4*(1), 252–267.

Tian, S., Liang, S., Qiao, K., Wang, F., Zhang, Y., & Chai, T. (2019). Co-expression of multiple heavy metal transporters changes the translocation, accumulation, and potential oxidative stress of Cd and Zn in rice (*Oryza sativa*). *Journal of Hazardous Materials, 380*, 120853.

Tollsten, L. & Muller, P. (1996). Volatile organic compounds emitted from beech leaves. *Phytochemistry, 43*, 759–762.

Toth, G., Hermann, T., Szatmári, G., & Pásztor, L. (2016). Maps of heavy metals in the soils of the European Union and proposed priority areas for detailed assessment. *Science of the Total Environment, 565*, 1054–1062.

USGS Minerals Yearbook (2013). http://minerals.usgs.gov/minerals/pubs/myb. html (Accessed 24 October, 2013).

Van De Mortel, J. E., Almar Villanueva, L., Schat, H., Kwekkeboom, J., Coughlan, S., Moerland, P. D., Ver Loren van Themaat, E., Koornneef, M., & Aarts, M. G. (2006). Large expression differences in genes for iron and zinc homeostasis, stress response, and lignin biosynthesis distinguish roots of Arabidopsis thaliana and the related metal hyperaccumulator *Thlaspi caerulescens. Plant Physiology, 142*(3), 1127–1147.

Van der Ent, A. J. M. M. J., Baker, A. J. M., Van Balgooy, M. M. J., & Tjoa, A. M. M. J. (2013). Ultramafic nickel laterites in Indonesia (Sulawesi, Halmahera): Mining, nickel hyperaccumulators and opportunities for phytomining. *Journal of Geochemical Exploration, 128*, 72–79.

Van der Zaal, B. J., Neuteboom, L. W., Pinas, J. E., Chardonnens, A. N., Schat, H., Verkleij, J. A., & Hooykaas, P. J. (1999). Overexpression of a novel Arabidopsis gene related to putative zinc-transporter genes from animals can lead to enhanced zinc resistance and accumulation. *Plant Physiology, 119*(3), 1047–1056.

Verbruggen, N., Hermans, C., & Schat, H. (2009). Molecular mechanisms of metal hyperaccumulation in plants. *New Phytologist, 181*(4), 759–776.

Wang, G., Wang, L., Ma, F., You, Y., Wang, Y., & Yang, D. (2020). Integration of earthworms and arbuscular mycorrhizal fungi into phytoremediation of cadmium-contaminated soil by *Solanum nigrum* L. *Journal of Hazardous Materials, 389*, 121873. https://doi.org/10.1016/j.jhazmat.2019.121873.

Wang, Y., Narayanan, M., Shi, X., Chen, X., Li, Z., Natarajan, D., & Ma, Y. (2022). Plant growth-promoting bacteria in metal-contaminated soil: Current perspectives on remediation mechanisms. *Frontiers in Microbiology, 13*, 966226. https://doi.org/10.3389/fmicb.2022.966226.

Weber, M., Trampczynska, A., & Clemens, S. (2006). Comparative transcriptome analysis of toxic metal responses in Arabidopsis thaliana and the Cd2+- hypertolerant facultative metallophye *Arabidopsis halleri. Plant, Cell & Environment, 29*(5), 950–963. https://doi.org/10.1111/j.1365-3040.2005. 01479.x.

Whiting, S. N., Leake, J. R., McGrath, S. P., & Baker, A. J. (2000). Positive responses to Zn and Cd by roots of the Zn and Cd hyperaccumulator *Thlaspi caerulescens*. *The New Phytologist, 145*(2), 199–210.

Williams, L. E., Pittman, J. K., & Hall, J. L. (2000). Emerging mechanisms for heavy metal transport in plants. *Biochimica et Biophysica Acta (BBA)-Biomembranes, 1465*(1–2), 104–126.

Wu, S., Zhou, S., Li, X., Johnson, W. C., Zhang, H., & Shi, J. (2010). Heavy-metal accumulation trends in Yixing, China: An area of rapid economic development. *Environmental Earth Sciences, 61*, 79–86.

Wuana, R. A. & Okieimen, F. E. (2011). Heavy metals in contaminated soils: A review of sources, chemistry, risks and best available strategies for remediation. *International Scholarly Research Notices, 2011*(11), 402647. https://doi.org/10.5402/2011/402647.

Xiao-e, Y., Xin-Xian, L., & Wu-zhong, N. (2002). Physiological and molecular mechanisms of heavy metal uptake by hyperaccumulting plants. *Journal of Plant Nutrition and Fertilizers, 8*(1), 8–15.

Yabe, J., Ishizuka, M., & Umemura, T. (2010). Current level of heavy metal pollution in Africa. *Journal of Veterinary Medical Science, 72*, 1257–1263.

Yadav, R., Arora, P., Kumar, S., & Chaudhury, A. (2010). Perspectives for genetic engineering of poplars for enhanced phytoremediation abilities. *Ecotoxicology, 19*, 1574–1588.

Yan, A., Wang, Y., Tan, S. N., Mohd Yusof, M. L., Ghosh, S., & Chen, Z. (2020). Phytoremediation: A promising approach for revegetation of heavy metal-polluted land. *Frontiers in Plant Science, 11*, 359.

Yang, M., Jawitz, J. W., & Lee, M. (2015). Uranium and cesium accumulation in bean (*Phaseolus vulgaris* L. var. vulgaris) and its potential for uranium rhizofltration. *Journal of Environmental Radioactivity, 140*, 42–49.

Yang, W., Gu, J., Zhou, H., Huang, F., Yuan, T., Zhang, J., Wang, S., Sun, Z., Yi, H., & Liao, B. (2020). Effect of three Napier grass varieties on phytoextraction of Cd- and Zn-contaminated cultivated soil under mowing and their safe utilization. *Environmental Science and Pollution Research, 27*, 16134–16144.

Yang, X., Feng, Y., He, Z., & Stoffella, P. J. (2005). Molecular mechanisms of heavy metal hyperaccumulation and phytoremediation. *Journal of Trace Elements in Medicine and Biology, 18*(4), 339–353.

Yuan, H. M., Xu, H. H., Liu, W. C., & Lu, Y. T. (2013). Copper regulates primary root elongation through PIN1-mediated auxin redistribution. *Plant and Cell Physiology, 54*(5), 766–778.

Zhang, G., Fukami, M., & Sekimoto, H. (2002). Influence of cadmium on mineral concentrations and yield components in wheat genotypes differing in Cd tolerance at seedling stage. *Field Crops Research, 77*(2–3), 93–98.

Zhu, X. F., Wang, Z. W., Dong, F., Lei, G. J., Shi, Y. Z., Li, G. X., & Zheng, S. J. (2013). Exogenous auxin alleviates cadmium toxicity in *Arabidopsis thaliana* by stimulating synthesis of hemicellulose 1 and increasing the cadmium fixation capacity of root cell walls. *Journal of Hazardous Materials, 263,* 398–403.

Zobel, R. W., Kinraide, T. B., & Baligar, V. C. (2007). Fine root diameters can change in response to changes in nutrient concentrations. *Plant and Soil, 297,* 243–254.

Index